VITAMIN C
Its Chemistry and Biochemistry

Royal Society of Chemistry Paperbacks

Royal Society of Chemistry Paperbacks are a series of inexpensive texts suitable for teachers and students and giving a clear, readable introduction to selected topics in chemistry. They should also appeal to the general chemist. For further information on selected titles contact:

Sales and Promotion Department
The Royal Society of Chemistry
Thomas Graham House
The Science Park
Milton Road
Cambridge CB4 4WF

Titles Available

Water *by Felix Franks*
Analysis – What Analytical Chemists Do *by Julian Tyson*
Basic Principles of Colloid Science *by D. H. Everett*
Food – The Chemistry of Its Components (Second Edition)
by T. P. Coultate
The Chemistry of Polymers *by J. W. Nicholson*
Vitamin C – Its Chemistry and Biochemistry
by M. B. Davies, J. Austin, and D. A. Partridge

How to Obtain RSC Paperbacks

Existing titles may be obtained from the address below. Future titles may be obtained immediately on publication by placing a standing order for RSC Paperbacks. All orders should be addressed to:

The Royal Society of Chemistry
Distribution Centre
Blackhorse Road
Letchworth
Herts. SG6 1HN

Telephone: Letchworth (0462) 672555
Telex: 825372

Royal Society of Chemistry Paperbacks

VITAMIN C
Its Chemistry and Biochemistry

MICHAEL B. DAVIES
JOHN AUSTIN
DAVID A. PARTRIDGE
Department of Applied Science
Anglia Polytechnic, Cambridge

ISBN 0-85186-333-7

A catalogue record for this book is available from the British Library

Published by The Royal Society of Chemistry,
Thomas Graham House, Science Park, Cambridge
CB4 4WF

Typeset by Servis Filmsetting Limited, Manchester
and printed by The Bath Press, Lower Bristol Road, Bath

Foreword
by
Professor M. Stacey CBE, FRS, DSc., C.Chem., FRSC

The steady output of publications on the chemistry, biochemistry, and medical applications of the still mysterious vitamin C has kept alive my memory of those exciting days of 60 years ago. It was the time of change from the 'test-tube' chemistry of organic chemical structure so brilliantly practised by Sir Robert Robinson and his pupils to the application of physical techniques such as ultraviolet spectroscopy, optical rotatory dispersion and X-ray crystallography. Biochemistry was just beginning to emerge as a separate science due to men like Professors Gowland Hopkins, A. Harden, and Harold Raistrick.

It is not always appreciated that Professor Haworth owed much of his success in the carbohydrate field to the development of microanalysis by Pregl. In 1925 H. D. K. Drew was sent from Birmingham to Prague to learn the techniques and to bring back the famous Kuhlmann balance and other apparatus. 5 milligrams was all that was now needed for the analysis of a methylated sugar and this was of enormous help with the meagre supplies of vitamin C. Another invention of Haworth was the formation of a team, which he called a 'syndicate' whereby a crash programme could be carried out. Each researcher had to drop his own research topic and concentrate on a particular stage in the synthesis or degradation! The tension in the laboratory was due to the fact that the Professors knew that their rivals in Europe, notably Professors Karrer, Reichstein, and Micheel, had material and were coming very close to the correct structure. Haworth anticipated the potential value of X-ray crystallography and in 1928 had recruited C. G. Cox (now Sir Gordon) to work on carbohydrate structures. With vitamin C, Cox quickly showed that the carbon and oxygen atoms lay in the same plane, confirming that it was, indeed, from his earlier studies, a carbohydrate.

In the summer of 1933 I spent six weeks in the laboratories of BDH in Graham Street, London, exploring with their chemists the possibility of an industrial-scale process for the vitamin and indeed the half kilogram we made is still in our museum! However, along came the remarkable five-stage synthesis of Reichstein and Grüssner, where the key stage is the biological oxidation of D-sorbitol to L-sorbose. One must pay tribute to the brilliant chemists and chemical engineers of the Roche Company who put this process on to the enormous industrial scale and where the process still holds its own. A few years ago I had the great thrill of seeing a silo holding 30 tons of glucose and mountains of 2 cwt polythene bags of beautifully crystalline vitamin C awaiting export. Production continues to grow.

For this book the authors have ranged far and wide to bring up to date all recent advances concerning the vitamin. The chapter on the history of scurvy is fascinating. One wonders why the conquest of this disease took so long when the answer was 'staring the doctors in the face', for so little of the vitamin is needed to prevent the disease. All the chapters are well researched and structures set forth in a clear manner. Of great importance and interest is the study of the medical aspects.

If all the claims for the biological functions and curative properties of this simple yet mysterious molecule are true, then surely we have discovered something near to the Elixir of Life so long sought by the ancient alchemists! However, the authors gave us a warning that in some cases the vitamin can be harmful, especially in megadoses. The biological systems of humans do show wide variations and what is sauce for the goose is not always sauce for the gander!

It has been a pleasure to read the manuscript and I warmly commend it, especially to students seeking inspiration.

Maurice Stacey
Birmingham 1991

Acknowledgements

No book is ever written without the combined help of a large number of people. We would like to thank our wives and families for putting up with us during the writing and Margaret Haw, who typed the manuscript, for her patience in deciphering our almost illegible scribbles and for not complaining as we continued to alter the manuscript even at the very last moment. We would like to acknowledge the help of those who gave their time to read the manuscript and provide very valuable comments: Dr Roger Mortimer, Dr Ian Fiddes and Paul Lowing; and the help of our colleagues at work, particularly John Hudson for valuable discussions on the historical aspects. We are also grateful to Dr Edwin Constable for valuable advice on the preparation of diagrams.

Special mention must be given to Professor Maurice Stacey, CBE, FRS, whose first-hand knowledge of the events surrounding aspects of vitamin C in the 1930s provided us with a fascinating insight into developments and personalities at that time.

In addition, the advice and assistance of the following individuals is gratefully appreciated: Peter Carter (Schering Ltd), Clive Conduit and David Godfrey (Roche Products Ltd), Paul Skelton (University Chemical Laboratories, Cambridge) and Bruce Robertson and Caroline Robinson (Media Production, Anglia Polytechnic).

Finally, we must acknowledge the countless workers in vitamin C chemistry and biochemistry over the past century or so and the many authors of the hundreds of papers and books which we have consulted in our work on vitamin C chemistry and in the preparation of this book.

If we have forgotten to mention anyone who has helped we apologise, but are, nevertheless, still grateful. Any mistakes in this book are of course entirely due to the authors, though we have striven to keep them to an absolute minimum!

Contents

Chapter 1

Introduction

Everyone has heard of vitamin C. There can be few simple organic molecules which have excited such universal interest. At least part of the reason for this has been the general interest in the beneficial effects of all vitamins and other trace substances on human health which has developed in recent years along with concern on the effects of other substances, particularly additives, on those who consume food containing them. We know that vitamins are essential to our well-being and because of this they have excited an interest and curiosity which has resulted in many of them being attributed with disease-healing and health-giving properties which they could not possibly have. Vitamin C has itself been said to have almost magical properties by some writers and it is useful to get a picture of the chemistry and biochemistry of this enigmatic compound.

Vitamin C is different. It is different from the other vitamins and we shall see in the course of this book that its chemistry and biochemistry single it out amongst molecules in many important ways. Vitamin C is ubiquitous. It is found throughout the plant and animal kingdoms, where its roles are often not known or are poorly understood. The synthetic vitamin is very widely used as a food additive and therefore has an E number (E300). However, unlike many other additives, few people would object to its presence in foods. There is no doubt that its anti-oxidant properties confer stability on foods to which it has been added.

Vitamin C has been the subject of frequent controversy, even before its nature had been established. Its role (as a constituent of fruits and vegetables) in the cure and prevention of scurvy was widely debated for hundreds of years. Its very existence was doubted by many even as recently as the the beginning of the twentieth century. There were quarrels over who was the first to discover it. Even today there is much controversy about the exact role of the vitamin in human health and there is not even agreement over the amount of the vitamin which

1

needs to be consumed for optimum well being, with various authorities recommending amounts varying from 30 mg to 10 g per day. The role in the relief of cold symptoms, in the improvement of quality of life for cancer patients and in other medical areas are all topics for intense discussion. The biochemistry of L-ascorbic acid in mammals is very poorly understood, so that it is not even clear what the biochemical role of the vitamin is in such systems. Although the chemical structure of L-ascorbic acid has been unequivocally established by single crystal X-ray diffraction, the structure of its very important two-electron oxidation product, dehydroascorbic acid, has not been finally established, since it has not yet proved possible to isolate crystals, or indeed the pure compound, as a solid.

Vitamin C is chemically the simplest of the vitamins and for this reason was among the first to be isolated, characterised, and purified and to have its structure determined. More vitamin C is produced industrially than any other vitamin, or indeed all the other vitamins put together. It is one of the few pure chemical compounds which is taken routinely by human beings in gram quantities (a possible challenger is sugar). It appears to have no harmful effects even in these large amounts and it is a medicine which it is a pleasure to take, especially in the form of fruit or vegetables.

It may be thought that the chemistry of this simple molecule would no longer hold any surprises after the vast amount of research that has been carried out over the years. However, conferences on aspects of vitamin C chemistry still attract large numbers of workers in the field and new aspects of the chemistry are always being revealed. The reason for the continued interest in the chemistry of L-ascorbic acid lies in the fact that despite it being such a simple molecule, its ene-diol structure provides it with a highly complex chemistry. Thus it has a very complicated redox chemistry involving comparatively stable radical intermediates which is heavily modified by the acidic properties of the molecule. It has been known for many years that L-ascorbic acid is easily oxidised by dioxygen. Although the first product of this process is dehydroascorbic acid, which still has antiscorbutic properties, the further oxidation by oxygen produces compounds which are not readily converted back to L-ascorbic acid, and the vitamin is effectively destroyed. The mechanisms of the reactions involved are still largely unknown, although they have been widely studied. There has been much recent work on the interactions of vitamin C with metal ions, particularly transition metal ions. This has unearthed a rich vein of chemistry involving L-ascorbic acid as both a redox companion and as a complexing agent; indeed the reaction of L-ascorbic acid with oxygen and other oxidising agents is catalysed by transition metal ions, especially copper(II), so that sometimes solutions are stabilised by the addition of EDTA, which complexes the metal ions and arrests the catalysis. It appears that vitamin C may not always act alone in its biochemical processes, but may act synergistically with other substances, of which vitamin E may be a typical example.

Research into vitamin C chemistry appears to have reached a kind of steady state. In the years 1969, 1979, and 1989, there were about the same number of papers published each year on aspects of the chemistry of L-ascorbic acid. Thus the extent of the work on this compound has been remarkably constant over the past twenty years and there is no sign of a diminution of interest yet.

The development of analytical techniques to detect and determine vitamin C has been crucial in the understanding of the presence and stability of the compounds in nature. At first, biological techniques were used and these gradually became replaced by chemical methods which were more sensitive, more selective, and easier to carry out. Today the analysis is largely centred around the use of high-performance liquid chromatography and there are many successful methods available. However, the technique is still limited by the fact that detection of dehydroascorbic acid in the presence of L-ascorbic acid still has a comparatively low sensitivity by virtually all detection techniques. This makes the determination of amounts of dehydro-ascorbic acid in plants and animals much more difficult than for L-ascorbic acid. The same applies to further oxidation products and there is a need for further work in the development of detection methods for these compounds and to investigate the extent and kinetics of oxidation of L-ascorbic acid in fruits and vegetables by determining the amounts of all oxidation products and the oxidation pathways by which these products are formed.

We are familiar with the use of ascorbic acid in pharmaceutical preparations. Although it is often found as the pure compound, in most cases it is present with a wide variety of other substances, which are often present simply to make the vitamin more palatable. However, much ascorbic acid is used for less well-known purposes. Much research has been carried out on the effects of ascorbic acid on various aspects of plant growth. It has been found to have effects upon germination and root growth. Spraying with ascorbic acid has been found to be effective in the protection of plants against the worst effects of ozone in the atmosphere produced by photolytic action on polluted air, particularly in big cities. Most domestic and farm animals are able to synthesise their own vitamin C, but even so this is sometimes reinforced by additional ascorbic acid. Fish are unable to synthesise ascorbic acid and the results of a vitamin C deficiency in fish are collectively known as 'broken back syndrome'. Thus, fish with a low dietary intake of vitamin C commonly suffer from distortions of the vertebral column, impaired collagen production, poor growth, and other symptoms. The widespread increase in aquaculture has meant that large amounts of ascorbic acid are used in the breeding and rearing of fish.

When L-ascorbic acid is used in the food industry, we can broadly divide the applications into two categories.

(1)   It is frequently used as an additive to foods where it enhances the nutritious qualities. It may be added to restore loss of vitamin due to the food processing or to increase the natural amount of the vitamin present. In either case the term nutrification has been used to describe its addition. Thus, L-ascorbic acid is added to fruit juice to fortify that which is naturally present or it may be added to artificial fruit drinks to improve taste and the nutritiousness of the drink.

(2)   L-Ascorbic acid may also be used as a food additive in circumstances where it is not expected to provide any increase in the nutritious nature of the food, but where it is present to prevent oxidation, as a preservative, to increase acidity, as a stabiliser, or as a flour improver. It is very often used as an additive for all these purposes.

Red meat has its characteristic colour because of the myoglobin and similar iron complexes which are present. This red colour is enhanced by the addition of nitrite as part of the curing process. This is due to the reaction of nitrogen monoxide with the myoglobin. The addition of ascorbic acid to meat also improves colour, flavour and odour, as well as lowering the amount of nitrite which has to be used in curing. Almost as a side effect it has been found that ascorbic acid, alone or in co-operation with tocopherol when used in meat curing, inhibits the formation of nitroso compounds, which are believed to be carcinogenic, while not interfering significantly with the inhibition by nitrite of the very dangerous *Clostridium botulinum* micro-organism.

L-Ascorbic acid is very widely used in bread baking, where it is present as a 'flour improver'. In practice, this means that the addition of L-ascorbic acid improves the bread texture and the size of the resulting loaf, the dough has greater elasticity, increased gas retention, and improved water absorption. Furthermore the addition means that storage time can be saved, since flour to which L-ascorbic acid has been added behaves rather like flour which has been matured over time. It is comforting to know that the products of the decomposition of L-ascorbic acid in bread making are carbon dioxide, L-threonic acid, and 2,3-diketogulonic acid and NOT oxalic acid! In many countries, it is the only flour improver which is allowed. Where others are used, it is interesting that they are all oxidising agents such as potassium bromate. L-Ascorbic acid is the only reducing agent used for such purposes. The exact mode of action of the vitamin C is still a mystery. It is clear that during the dough mixing operation all the L-ascorbic acid is converted into dehydroascorbic acid and this remains stable in the mixture. Perhaps the best-known process using L-ascorbic acid as an additive in bread making is the so-called Chorleywood process.

Some countries have a legal maximum for the amount of L-ascorbic acid used as a flour improver. These may be as high as 200 mg kg^{-1} in Canada, or as low as 20 mg kg^{-1} in Uruguay. Many other countries, however, such as the United Kingdom leave the amount added to Good Manufacturing Practice.

Other areas where L-ascorbic acid has found uses in industrial processes are, for example, in polymerisation reactions, in photographic developing and printing, in metal technology, and even in intravaginal contraceptives. Most of these applications involve the use of the reducing properties of L-ascorbic acid in some way.

L-Ascorbic acid is found all over the plant world, often in quite large quantities and distributed throughout the plant. The biochemistry of vitamin C in plants is very poorly understood. A view which seems to be accepted generally is that in some way L-ascorbic acid is merely a secondary product of plant metabolism. It seems curious that such a ubiquitous compound in plants should be there almost incidentally as a by-product of other processes, though it is fortunate for those creatures that have lost their ability to synthesise the vitamin that it is so!

There is evidence that tartaric acid in grapes has L-ascorbic acid as a major precursor. Thus, immature grapes fed with L-ascorbic acid labelled with ^{14}C at C-1 were found to have 72% of the ^{14}C in the carboxyl carbon of tartaric acid. The indications are that in L-ascorbic acid C-1 to C-4 end up as tartaric acid; indeed, if C-6 is labelled with ^{14}C none of it appears in the tartaric acid. However, the process in vines is by no means simple and depends on a number of factors, including the extent of development of the plant, and is different for different parts of the plant. In a similar way, oxalate has been found as a possible product of biosynthetic routes in some plants such as the geranium. Many plants are capable of accumulating comparatively large amounts of oxalic acid, sometimes even as crystals of calcium oxalate. Perhaps the best known oxalate accumulator is rhubarb. It seems likely that in these plants as well, the oxalic acid produced has L-ascorbic acid as a precursor. The biosynthesis of L-ascorbic acid itself from D-glucose in plants occurs by a process quite different from the route believed to take place in those animals which are able to synthesise it. In plants, the biosynthetic route is thought to involve the oxidation of C-1 of glucose, epimerisation of C-5, and retention of the C-6 hydroxymethyl group.

The process of photosynthesis, which is so vitally important to the existence of virtually all life of Earth, is now known to be an extraordinarily complex process which is still only poorly understood. It is known, however, that although dioxygen is essential to the development of all life, nevertheless high concentrations have an adverse effect on a number of important biochemical processes including photosynthesis. In this case, high concentrations of dioxygen inhibit the development of chloroplasts. In the process of illumination of chloroplasts, damaging oxygen-containing species such as hydrogen peroxide and singlet oxygen may be formed. It is suggested that the damaging effects of these species is limited by the presence of L-ascorbic acid in plants, which acts to suppress these highly oxidising molecules.

The arguments will continue over the biochemical role of L-ascorbic acid in animals, over the reasons for its large concentrations in plants, and over who was the first to discover and isolate it. Interest in its chemistry will continue and there is certainly much to discover. However, for most people, its role in medicine will remain the most interesting and controversial area. There is much conflicting evidence concerning the amount we should be taking. Does it help to relieve the symptoms of the common cold? Does it have a role in cancer therapy? Perhaps the best position to take up is to stay with what is known for certain. Vitamin C is clearly vital to the production of collagen and is important in wound healing. For this reason alone it is probably wise to ensure that we receive amounts rather higher than those recommended for the suppression of scurvy. Although there is not much evidence for serious effects on taking very large amounts, perhaps it is unwise to take tens of grams at a time of any compound. We are often asked how much should be taken daily. The above is the nearest we would give to an answer!

Chapter 2

History of Vitamin C and Its Role in the Prevention and Cure of Scurvy

Nothing emphasises the importance of vitamin C to human beings more than the effect of being without it for a relatively short time. Just a few months' deprivation produces the particularly unpleasant and ultimately fatal disease, scurvy. It is hard for us today to appreciate the fear with which this mysterious disease was regarded, particularly by seafarers in the middle ages. Although it was only one disease among many which afflicted sailors, it seemed to flare up for no apparent reason, particularly on long sea voyages, which were becoming more and more common from the fourteenth century. It was not uncommon to lose more than half the crew on such a journey. Vasco da Gama lost half his complement of men when he first rounded the Cape between 1497 and 1499 and scurvy continued to take its toll of sea travellers for four hundred years after this time.

What is this disease? A 'textbook' definition would be something like: *A disease which produces haemorrhaging into tissues, bleeding gums, loose teeth, anaemia and general weakness.* However, contemporary descriptions of individual cases bring home to us the unpleasantness of scurvy. Thus Thomas Stevens wrote from a ship travelling from Lisbon to Goa in 1579:

'. . . their gums wax great, and swell, . . . their legs swell, and all the body becometh sore, and so benumbed, that they can not stir hand nor foot, and they die for weakness, or fall into fluxes and agues, and die thereby. . . .'

Father Antonia de la Ascension in 1602 wrote in his diary while on expedition along the coast of California:

'. . . The first symptom they notice is a pain in the whole body which makes it sensitive to touch . . . all the body, especially from the waist down becomes covered with purple spots . . . The sensitiveness of the

bodies of these sick people is so great that . . . the best aid which can be
rendered them is not even to touch the bedclothes . . . the upper and
~~lower gums of the mouth in the inside of the mouth and outside the teeth,~~
become swollen to such a size that neither teeth nor the molars can be
brought together. The teeth become so loose and without support that
they move while moving the head . . . their natural vigour fails them and
they die all of a sudden, while talking.'

We now know that scurvy is a deficiency disease and it must have
existed throughout all history. However, it was really the very long sea
journeys of the early Middle Ages which drew attention to it, though
any disasters which affected the food supply, such as sieges, would also
result in an upsurge of scurvy. It would appear, however, that we can
identify at least three major factors which contributed to the failure to
recognise the cure for scurvy.

(1) The Lack of Communication in Ancient Medicine

It seems likely that the fact that consumption of citrus fruits could
cure scurvy was known or suggested very early on. Winslow and
Duran-Reynals have quoted a thirteenth century Spanish medical
tract recommending orange and lemon juice as being beneficial to
scurvy sufferers. Similar statements are to be found over the next four
centuries, but were not widely available, nor does there appear to have
been a custom of widely reading the works of medical experts elsewhere
in the world, even when they were available. It took an incredibly long
time for something which seems obvious today to become accepted by
the medical establishment.

(2) The Instability of Vitamin C

We will examine the chemistry of the oxidation of vitamin C in later
chapters. It has been known for a very long time that the antiscorbutic
effects of fresh fruits and vegetables diminish after the time that they
have been harvested. This is because of loss of vitamin C due to a
variety of oxidative processes involving oxygen from the air. This
contributed to the confusion over the efficacy of fruit, vegetables and
fruit juices in the prevention and cure of scurvy, even as late as the
beginning of the twentieth century.

(3) Distribution of Scurvy Amongst the Population

Scurvy was perceived to be a disease of a relatively narrow section of
the population and was thought not to afflict the rich and famous.
However, we shall see later that there is evidence of its appearance
during the winter months among the aristocracy and indeed it may
even have afflicted members of the the Royal Household, though it was
not recognised at the time. It tended not to be a disease of cities. This

and a lack of understanding of the pathology meant that this was not a 'fashionable' disease that captured the interest and the imagination of the medical establishment.

One of the earliest accounts of cures of scurvy is to be found in Hakluyt's 'Principal Navigations' which was published in 1600. This was referring to events which took place on Jacques Cartier's expedition to Newfoundland in 1535.

'Some did lose all their strength . . . others also had all their skins spotted with spots of blood of a purple colour: then did it ascend to their ankles, thighs, shoulders, arms and necks. Their mouths became stinking, their gums so rotten that all the flesh did rot off, even to the roots of the teeth, which almost all fall out. Our Captain, considering our estate and how sickness was increased and hot amongst us, one day went forth from the fort, and walking upon the ice saw a troop of those countrymen coming from Stadacona, amongst which was Domagaia, who not ten or twelve days before, had been very sick with that disease, and had his knees swollen as big as a child of two years old, all his sinews shrunk together, his teeth spoiled, his gums rotten and stinking. Our Captain, seeing him whole and sound, was marvellous glad, hoping to understand how he had healed himself, to the end he might ease and help his own men. As soon as they came near he asked Domagaia how he had healed himself: he answered that he had taken the juice and the sap of the leaves of a certain tree, and had with these healed himself. Then our Captain asked him if any were to had thereabout . . . Domagaia straight sent two women to fetch some of it, who brought ten or twelve branches of it, and then showed how to use it and that is, to take the bark and leaves and boil them together, then to drink the said decoction every other day . . . The tree in their language is called Ammeda or Hannedew, this thought to be the Sassafras tree. Our Captain presently caused some of that drink to be made for his men, but there were none durst taste of it, except one or two, who ventured drinking of it: others seeing did the same, and presently recovered their health and were delivered of their sickness, and with this drink were clean healed. After this medicine was found and proved to be true, there was much strife about it, who should be the first to take it, that they were ready to kill one another. A tree as big as any oak in France was spoiled and lopped bare, and occupied all for 5 or 6 days and it wrought so well, that if all the physicians in Montpelier or Lovain had been there with all the drugs of Alexandria, they would not alone so used in one year, as that tree did in six days, for it did so prevail, that as many as used of it, by the Grace of God recovered their health.'

The cases which have been quoted and described here are clear-cut. The major problem for any ship-board physician was that the sailors were subject to a large number of different diseases arising from a variety of causes. These were usually poorly described and therefore it was often difficult or impossible to give a definite diagnosis or to distinguish between different diseases. The result was that many cases of scurvy went undiagnosed thus complicating the treatment.

One of the first references in English to the disease of scurvy was in Hakluyt's 'Principall Navigations', which was first published in 1589, where he records that in a 1582 expedition returning from the Straits of Magellan two men died of 'skurvie' when nearly home.

Most of the early long-distance exploration by sea was carried out by the Spanish and Portuguese. However, British exploration began in earnest in the sixteenth century, epitomised by Drake's circumnavigation of the world in 1577–1579. Like the Portuguese and Spanish sailors, the British were plagued by scurvy. Again there are passing references to the effectiveness of various fruits and herbs in the treatment of the disease. Thus the journals of the voyage of circumnavigation undertaken by Thomas Cavendish in 1586 refer frequently to the importance of fresh fruit and there were relatively few cases of scurvy on this voyage. We can get an idea of the length of time it took to encounter scurvy on a sea voyage from the account of his voyage in 1593, provided by Sir Richard Hawkins. They began the voyage in April and by the time they reached the Equator in August, some of the crew had gone down with scurvy. By October, there were only four healthy individuals amongst the crew. The situation was then saved in Brazil, a Portuguese colony at that time, where oranges and lemons were purchased. Although Hawkins recommended oranges and lemons as a cure for the disease, he nevertheless considered that drinking dilute (very) sulphuric acid was helpful, but most of all 'air of the land', considering that the sea was the natural place for fishes and land the natural place for man!

Although by no means generally accepted in Britain, it is clear that the notion that fruit juices helped in the treatment of scurvy was gradually becoming better known. The journal of Sir James Lancaster's voyage to Sumatra in 1601 is quite unequivocal about the value of oranges and lemons:

> '... the reason why the general's men stood better in health was this; he brought to sea with him bottles of the juice of lemons, which he gave to each one, as long as it would last, three spoonfuls every morning ... By this means the general cured many of his men. ...'

Later in the journey he repeated the process by calling at the Bay of Antongile to 'refresh our men with oranges and lemons, to clear ourselves of the diseases'.

Among the victuals recommended to be carried on expeditions

funded by the East India Company, lemon water was included. Even in 1626 some argued that tamarind was of more value than other fresh fruit in the treatment of scurvy, though fruit juice continued to be supplied to ships of the East India Company.

SUGGESTED CAUSES OF SCURVY

It is worth pausing at this point in the consideration of the development of a cure for scurvy to consider what people believed to be the cause of the disease. This is not intended to be an exhaustive treatment of this topic, but to highlight some of the early theories.

One of the earliest suggestions was that scurvy was a disease of the spleen. Even in the late Middle Ages the ideas about the operation of the body and the causes of it going wrong were very different from those which we hold today, although as today they were believed to be absolutely correct, little having changed from the time of the Ancient Greeks, whose approach to medicine was held in the greatest respect. Hippocrates of Cos lived from 460 to 377 BC and was a contemporary of Socrates. The teachings of the Hippocratic School of Medicine were brought together as a collection of sixty texts which are known as the Hippocratic Corpus. Although much of the contents can be attributed to Hippocrates, there are nevertheless many contributions from other scholars including Hippocrates' son-in-law, Polybios. A characteristic of Greek medicine was that they only had a very hazy knowledge of the structure and functions of the various parts of the human body. They were particularly unclear about the nature and functions of the vital organs and, of course, knew nothing about the circulation of the blood. They were familiar with the fact that the human body secreted fluids. It was obvious to them that these had something to do with disease; excessive bleeding caused illness and ultimately death, a running nose is a feature of the common cold and other illnesses, people often vomit when they are ill, and so on. They therefore formulated a theory incorporating these fluids or 'humours'. The body was thought to contain and be influenced by four humours:

blood
black bile
yellow bile
phlegm

The balance between these humours governed the health of the individual. Each humour had associated with it a property:

Blood – sanguine or hopeful
Black bile – melancholy or gloomy
Yellow bile – choleric or irascible
Phlegm – phlegmatic or even tempered

These humours were considered to originate and be replenished by various organs: blood by the heart, black bile by the spleen, yellow bile by the liver, and phlegm by the brain. Lastly they had associated with them a physical property: blood – hot and wet, black bile – cold and dry, yellow bile – hot and dry, phlegm – hot and wet, as represented in Figure 2.1 These in their turn were considered to be related to the four elements, earth, air, water, and fire.

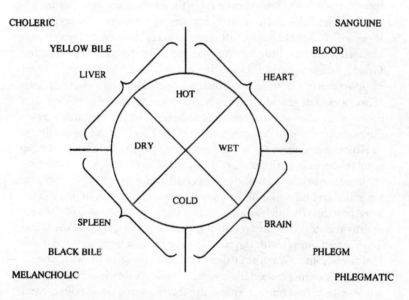

Figure 2.1 *The theory of the four humours*

Terms relating to the idea of these humours are of course still used today and we talk of people being 'melancholic', phlegmatic, bilious, sanguine, while someone who is ill tempered is called 'liverish'. It was believed that someone suffering from a disease such as scurvy had a swollen and hard spleen, so that its function was impeded. This meant that the black bile could not be purified in the recycling process and this would result in the symptoms of scurvy. Although we recognise today that this theory has no validity, it also suffers from the disadvantage that it does not suggest a cure, nor does it provide a process leading to a cure. However, since black bile was thought to be associated with melancholy, presumably by improving the disposition of the sufferer this would go some way towards curing the disease, and the provision of good clean air and pleasant surroundings, for example removing the unpleasantness of being crowded with other people on a small ship, would help to relieve the symptoms.

We take for granted today the value of chemistry in both the treatment of disease and the understanding of the causes of ill health. In the sixteenth and seventeenth centuries, chemistry as we know it today scarcely existed but one or two ideas about the interactions between substances which we would find acceptable today were gradually emerging from the darkness of alchemy. One of these was that some substances could be classified into acids and alkalies. There was no definition such as we have today, but it was known that there were acidic and alkaline substances which had corrosive effects upon the human body. It was assumed that the process of digestion, for example, involved an acid–alkali reaction. When the digestive process was not working properly, it would introduce, say, acid into the body and produce certain symptoms associated with an excess of acid. Such diseases could then be treated using alkaline preparations. Similarly, other types of digestive imbalance might result in an alkaline condition, when administration of acid would be prescribed. Scurvy was regarded as a mixture of these conditions, since the ulcers which were characteristic of the disease were thought to be an attribute of an acid condition, while the evil-smelling breath was a feature of an alkaline condition. Thus, both acid and alkaline medicines had to be used together with bleeding, which was generally regarded as being beneficial in a wide variety of ailments and was extensively used over several centuries, often with great enthusiasm! Indeed the use of bleeding in the treatment of disease continued right into the middle of the nineteenth century.

In many ways, scurvy was a disease of Empire. Those countries with territorial ambitions, such as Spain, Portugal, and Britain, had to develop powerful navies for exploration, to extend their empires and to defend their far-flung colonies. The British Navy began operating in this role at the end of the seventeenth century. From this time, scurvy began to be a major problem for the Navy. The war with Spain, which began in 1740, required the British Navy to extend itself over great distances in capturing Spanish ships bringing treasure from South America and her colonies elsewhere. The experiences were similar to those of Portuguese and Spanish sailors. After a few months at sea, men started to suffer the symptoms of scurvy. There are many references in this context to the efficacy of fresh fruit and vegetables in the treatment of the disease, but there was also much confusion and a widespread assumption that 'good' things such as fresh air, fresh meat, temperate weather were effective in its treatment. The standard medicine which was frequently prescribed for treatment was the so-called 'elixir of vitriol' which was essentially dilute sulphuric acid containing ethanol and sugar, made palatable by the addition of flavouring. It is easy for us to see today, with the benefit of hindsight, that against this background of confusion within which there was enough knowledge to provide a cure for scurvy, there was a need for a clear-cut scientific demonstration of the effectiveness of fresh fruit and vegetables. That is, what was needed was what we would call today a clinical trial. Something very close to this was to be provided by James Lind.

JAMES LIND

In the eighteenth century the normal way to become a surgeon was to be apprenticed. James Lind was born in 1716, in Edinburgh. However, his ancestors came from Dalry which was in Ayrshire. It is a remarkable coincidence that today Roche Products have a factory at Dalry manufacturing many thousands of tonnes of L-ascorbic acid every year. At the age of fifteen Lind became apprenticed to a local surgeon. He learned his trade in the same way as an apprentice to say a carpenter would become skilled, by watching, listening, and gradually being allowed to do more and more of the practical surgery. When he was 23 in 1739, he joined the Royal Navy and served as a surgeon's mate. He became a full surgeon at the age of 30 on *HMS Salisbury* and very soon he was faced with cases of scurvy with outbreaks occurring in 1746 and 1747. It was during the second outbreak that he set up what can only be described as the much-needed clinical trial to test the effectiveness of the various treatments of scurvy in use at that time.

He chose twelve men who had contracted scurvy and fed each the same diet throughout the day. He then applied six different treatments to the men, so that groups of two would each have identical treatment (Table 2.1). The result was that, although there were only enough oranges and lemons to carry out the treatment for six days, those treated in this way quickly recovered, to such an extent that they were able to resume normal duties. For the men provided with cider, at the end of two weeks 'the putrefaction of their gums, but especially their lassitude and weakness somewhat abated'. The other treatments had, as we now with the gift of hindsight would expect, little or no effect

Table 2.1 *A summary of James Lind's trial of various possible cures for scurvy*

Group	Treatment
1	One quart of cider
2	25 drops of elixir of vitriol three times a day
3	Two spoonsful of vinegar before meals
4	Half a pint of sea water
5	Two oranges and one lemon
6	A medicinal paste containing a variety of substances such as garlic, mustard seed

upon the condition of the men or the course of the disease. James Lind concluded from this experiment 'that oranges and lemons were the most effectual remedies for this distemper at sea'. In 1748 Lind left the Navy and went back to Edinburgh, where in 1750 he was elected a Fellow of Edinburgh's Royal College of Physicians. Then, in 1758 he became physician to the new naval hospital at Haslar in Gosport, near Portsmouth, which is still a working naval hospital today. The building of these naval hospitals was itself a response to the ever increasing problem of sickness in the Royal Navy at that time.

In 1753 Lind published his famous book entitled:

'A Treatise on the Scurvy in three parts containing An Enquiry into the Nature, Causes and Cure, of that Disease together with A Critical and Chronological View of what has been published on the Subject'.

A second edition was published in 1757.

The book gives an excellent account of the previous knowledge of scurvy, but his theories of the cause of the disease are difficult to interpret. Carpenter has pointed out that even in the 1953 symposium celebrating the two hundredth anniversary of the Treatise, no attempt was made to summarise the theory, which is based on ideas concerned with 'clogging' of the pores of the skin, which were themselves considered the major means of getting rid of unwanted 'humours and vapours'.

Lind made recommendations concerning what antiscorbutic victuals should be carried on ships, but even he remained unconvinced that green vegetables and fruit could actually prevent scurvy. He certainly recommended that ships should carry an extensive store of green vegetables as well as oranges and lemons.

Nobody in authority listened to Lind, however. He did not seem to have influence in the right places and there was a considerable delay before Gilbert Blane, a man who did have the right influence, came along.

SIR GILBERT BLANE

The influence of the British Navy in the development of the means of treatment and prevention of scurvy was further enhanced by the appointment in 1781 of Gilbert Blane as Physician to the British Fleet by Admiral Rodney. Blane was born in 1749 and was a graduate of Edinburgh and Glasgow Medical Schools. After obtaining his M.D., he practised as a physician and gained a considerable reputation. This resulted in his friendship with Admiral Rodney. He spent ten months with the fleet before preparing and sending a memorandum to the Admiralty concerning the casualty rate in the fleet. In the year before, 1600 men died, but only 60 of these died as a result of enemy action. Indeed there were only 12 000 men in the fleet altogether! While there were many fatal diseases, there is no doubt that many of those deaths could be directly attributed to scurvy. Blane stated at the time:

'. . . Scurvy, one of the principal diseases with which seamen are afflicted, may be infallibly prevented, or cured, by vegetables and fruit, particularly oranges, lemons or limes'

The memorandum recommended that arrangements be made to carry fresh fruit '. . . every fifty oranges may be regarded as a hand to the fleet'.

Having returned to London with Admiral Rodney in 1782, Blane went back to the West Indies for a further two years and after that did not return to sea again.

But still the confusion continued. Despite Blane's unequivocal statement concerning the effectiveness of fresh fruit and despite the great weight of practical evidence in its favour, the Board of Sick and Hurt Seamen rejected this and other suggestions by physicians that the provision of fresh fruit would solve the scurvy problem. Other nutritional solutions continued to be championed. Beer, molasses, fresh bread, all found their proponents. In the nature of medical trials and particularly uncontrolled trials, these and other remedies sometimes produced apparent cures. Even Gilbert Blane on his return to civilian life, writing in 'Observations on the Diseases of Seamen', was convinced of the antiscorbutic activity of other foods, including, particularly, molasses.

It is fortunate for the history of the cure of scurvy that Gilbert Blane became physician to the Household of the Prince of Wales. This gave him a social status and friendships which allowed him to have his strong views about the effectiveness of fruit in the cure of scurvy listened to in the right places. He was now able, through his friendship

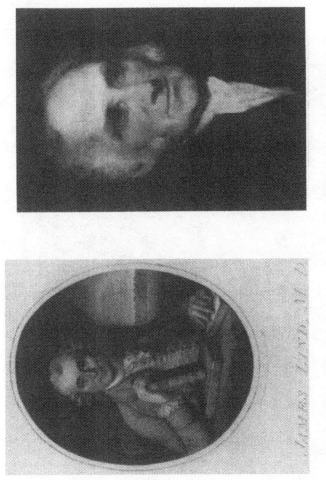

James Lind (1716–1794)
(Wellcome Institute Library, London)

Sir Gilbert Blane FRS (1749–1834)
(Royal College of Physicians, London)

E.L. Moss's watercolour of the daily distribution of lime juice during the 1875 Arctic expedition (Scott Polar Institute, Cambridge)

with Admiral Gardener, to get the Sick and Hurt Board to approve the idea of carrying lemon juice on a ship of the fleet (*HMS Suffolk*), which was sent to the East Indies. No men died of the scurvy on that voyage. The very few who contracted it were cured quickly by encouraging them to drink orange juice. When, in 1795, Blane became a commissioner to the Board, it authorised that lemon juice be a regular issue to the British Navy. The chosen allowance was three-quarters of an ounce per day, and from 1796 the incidence of scurvy dropped off dramatically. Scurvy was eliminated from the navy as a major problem from then on. This was two years after the death of James Lind and just about fifty years after his famous clinical trial which demonstrated the effectiveness of oranges and lemons in curing scurvy! Before his death in 1834, Gilbert Blane had been made a Fellow of the Royal Society and received a knighthood in 1812.

The scientific relationship between Lind and Blane is an interesting one. Lind died in 1795 with little or no recognition. He had retired from Haslar eleven years before. He had no influence on the victualing procedures of the Royal Navy. Blane, however, was able to recognise the most important aspects of Lind's and other people's work. Furthermore, because of his social standing and his social contacts, he was able to bring Lind's ideas to the right people and the appropriate committees, so that they could be applied to the nutrition policies of the Royal Navy. Both must be regarded highly in the history of the prevention and cure of scurvy. Many people lived who otherwise would have died because of the lives of Lind and Blane.

Largely because of Lind and Blane scurvy had been conquered at sea. This should be the end of the story, but in the nineteenth century events were happening on shore which would bring the curse of scurvy inland.

SCURVY ON LAND

In Northern Europe, during the winter months, fruit and vegetables have always been in short supply. It is likely that throughout history, the peoples of the British Isles must have been on the edge of suffering from scurvy or have suffered from what is now known as pre-clinical scurvy during the winter months. This situation may have been exacerbated in Christian countries by the very strict adherence to Lenten fasting in the six weeks before Easter. When Easter was early, this would have occurred just as fresh vegetables were becoming available after the winter. Furthermore the aristocratic people regarded vegetables as food which was fit only for the lower classes, because they grew in the 'dirty' soil. In some ways, therefore, the aristocracy were more prone to the possible ravages of scurvy as the winter wore on. In other words, scurvy on land was very much a seasonal disease, appearing in winter and beginning to disappear in spring. These facts and others have been pointed out by Susan Maclean Kybett in reasoning that the mystery illness suffered by Henry VIII in the last ten years of his life was probably scurvy, brought on by a large intake of food which did not contain any vitamins during the winter months and very strict Lenten fasting in the Spring. Kybett has pointed out that the English had a diet which relied heavily on the meat of birds and animals and, indeed, were afraid of eating fruit, which many believed caused skin eruptions and fevers. During the last ten years of his life, Henry became extremely fat, but suffered badly from leg ulcers, fevers, spots on his body, swings of mood, and other symptoms which are strongly reminiscent of those produced by scurvy. It is one of the curious features of scurvy that the patient may be well-fed in terms of quantity of food and hence be over-weight and yet be desperately ill from the disease.

Henry died on the 28th January 1547; the time of the year may be significant. It may be that scurvy made a major contribution to his death. Henry would not have been alone in suffering from the disease and there are records of other aristocratic people also having the symptoms, both in Henry's time and later. Thus, it seems very likely that James I and his wife both suffered from scurvy and no doubt it remained undiagnosed among the countless illnesses suffered by people in the Northern Hemisphere throughout the ages. It is not until the nineteenth century, however, that we begin to get a clear picture about the occurrence of scurvy on land as the focus moved away from the incidence of scurvy at sea.

Although historically potatoes were introduced into the Western World comparatively recently, the nutritional importance of the potato has frequently been underestimated. Many of us think of potatoes as the food which has to be cut down on in a diet to lose weight because of their high carbohydrate content. Carpenter and other authors have pointed out the importance of this humble vegetable in the prevention and cure of scurvy. Prisons in the nineteenth century were, next to hospitals, places to keep away from if you wished to remain healthy. At that time it was part of the philosophy of the treatment of criminal behaviour (and perhaps still is) that the wrong-doer should suffer, not just by loss of liberty, but by the deprivation of comfort and good food. As a direct result of this there were spasmodic outbreaks of scurvy in all British prisons at this time. Any prison governor in the early nineteenth century found himself in the difficult dilemma of trying to provide a diet which would be sufficient to prevent disease, but which would not be regarded by the general public as being too luxurious! Government cuts are not just a phenomenon of the late twentieth century and periodically governors were ordered to cut back even on the meagre rations which they supplied to those in their care. When this happened, one of the first things to be curtailed was potatoes. It became clear that this policy was actually allowing the development of scurvy in British prisons. Analysis of the food provision in prisons at that time, using the reports made by the Inspectors of Prisons, showed that, in gaols where potatoes were not a part of the diet, scurvy was to be found. As a result of this potatoes were recommended as part of every prisoner's diet. This was acceptable because the potato was a very cost-effective crop and cheaper than green vegetables. It was also for this reason that the potato had become a vital part of the diet of the poorer people of Ireland at this time.

The Summer of 1845 was particularly cold and wet even for the British Isles. Indeed the whole of Northern Europe was damp that year. During that Summer a disease began to attack potatoes which turned the leaves black and produced tubers which were deformed and discoloured. When they were stored, they rapidly rotted. Encouraged by the damp weather, the disease spread throughout the whole of Northern Europe. In 1845, virtually half the potato crop for that year was destroyed. What made things even worse was that the crop failed again in 1846. This resulted in a terrible famine, which hit Ireland particularly badly. Potatoes were a staple food, so that the first effect of the famine was widespread starvation. This weakened people, leaving them vulnerable to various diseases such as typhoid, and many consequently died. Things were also very bad in mainland Britain, but not as terrible as in Ireland, where the population was much more dependent upon the potato. Soon scurvy began to appear. Despite the evidence available about the use of potatoes in prisons, the appearance of scurvy came as a surprise to the medical establishment.

One of the problems with theories is that even when they are wrong, but are accepted, they guide and sometimes govern practice. One theory put forward at the time of the potato famine was that scurvy was caused by a lack of protein. This arose from an erroneous guess about what common feature was possessed by the various foods known, or believed, to have antiscorbutic properties. The result, when the theory was acted upon, was that the wrong foods were recommended in some prisons. The consequence of this was, of course, scurvy. Despite yet another wrong interpretation of the cause of scurvy, it at least now was gradually becoming accepted that there was some substance in the food which, if it were not present, would produce the disease. That is, in modern terminology, scurvy was a deficiency disease. A theory also prevalent at this time that scurvy was caused by the absence of potassium in some foods arose from similar thinking.

Meanwhile, a normal potato harvest was produced in 1848. That year, scurvy virtually disappeared. The famine had had the most appalling social consequences, particularly in Ireland, which are still felt today after nearly 150 years.

The supply of foods is always a major problem during wars and each of the skirmishes which spasmodically broke out during the nineteenth century had its crop of deaths from scurvy. In the American civil war, 25 percent of the captured soldiers who died in Andersonville prison died of scurvy. Scurvy killed many in the siege of Paris in 1870 and there was a very large outbreak of the disease during the Crimean war (1854–1856). There were the usual episodes of failure to deliver the right food to the right place during that war, as in all wars. Among the numerous diseases suffered by the soldiers looked after by Florence Nightingale, scurvy accounted for many deaths while the food supply situation was difficult.

Scurvy remained a problem even into the twentieth century and would continue to be so until the antiscorbutic essence in foodstuffs was revealed and isolated.

EXPLORATION OF THE ARCTIC AND ANTARCTIC

The last great areas of exploration at the end of the nineteenth century and the beginning of this century were the Arctic and Antarctic. There was much activity in the exploration of the Arctic throughout the nineteenth century. Typical of such expeditions was to journey around Cape Horn, up the American Coast and along the Northern coast of Alaska. Such a journey was undertaken by the ship *Investigator*, whose experiences have been described in detail by Carpenter. This and other ships became stuck in ice, often for several years. In many such cases, scurvy often struck, sometimes even when they had what was regarded as an antiscorbutic diet. Indeed, by the time of Scott's first expedition to the Antarctic, the idea that scurvy was caused by the primary decomposition products of meat, so-called ptomaines, had become widely accepted. Indeed the provisions which were carried on this expedition were carefully chosen to make sure that the levels of ptomaines remained at a minimum. This expedition set off in 1901, at the beginning of the twentieth century! By September 1902, there were the first signs of scurvy. At this stage, fruit juice and vegetables were provided. The men recovered. During an attempt to get as far south as possible that same year, Scott and Shackleton began to show the symptoms of scurvy from which they recovered when they returned. At the time, their recovery was attributed to the consumption of fresh seal meat.

In 1910, Scott set off on his famous and ill-fated expedition to the South Pole. Fruit juice was carried, but the men were not ordered to consume it. Fresh meat was carried in abundance at the beginning. There was no doubt that during the expedition, Lieutenant Edward Evans (later Admiral Lord Evans), contracted scurvy, from which he later recovered.There appears to be no evidence that scurvy contributed to the deaths of Scott, Evans, Oates, and Wilson, though it has been suggested that the circumstances of Evans's death were such that scurvy could have been a contributory factor. Such a suggestion at that time would have been regarded as scandalous.

SCURVY IN CHILDREN

In the second half of the nineteenth century, there was an increasing tendency to wean children away from mother's milk early. The milk was replaced by rusks and 'condensed milk'. Many such children belonged to wealthy families who would be shocked to hear that their offspring were prone to scurvy as a result of a poor diet. Children with scurvy suffered great pain and couldn't bare to have anyone touch their limbs. A little girl of just under two years old under the care of Dr Thomas Barlow, who was registrar at the Sick Children's Hospital, died in March 1874. A *post mortem* examination indicated that the entire periosteum of the femur was separated from the bone and the space between was filled with a blood clot. Similar effects were observed with other bones. This was diagnosed as being due to scurvy and the disease of infantile scurvy became frequently known as 'Barlow's Disease'. Many cases were found in Britain, America, and Germany, frequently superimposed on rickets and sometimes mistaken for congenital syphilis. It is interesting that one of the most obvious features of scurvy, bleeding gums, was not present in many cases of infantile scurvy because of the absence of teeth. Barlow treated children with scurvy with raw minced beef, fresh cow's milk, and orange juice and an improvement was usually observed in a very short time.

Much of the failure to conquer scurvy completely, even at the beginning of this century, was due to the continuing reluctance, despite the evidence, to recognise it as a deficiency disease. Until the common factor which the various antiscorbutic foods possessed could be identified, isolated, and shown to be effective in the cure of scurvy, the arguments about the causes of the disease, and which foods were active against it, would remain. More than a quarter of the twentieth century was to pass before this was to occur. However, during these twenty-five years, scientists were gradually unravelling the mystery of the elusive antiscorbutic factor.

THE EVENTS LEADING UP TO THE DISCOVERY OF VITAMIN C

It was well into the first decade of the twentieth century before it was finally recognised that there was a substance or substances in fruit and vegetables which was essential to the well-being of humans. One major problem in the investigation of scurvy was that it is not possible to induce the disease in many laboratory animals such as rats. We now know that the reason for this is that these animals are very proficient in the biosynthesis of their own vitamin C. It was in 1907 that Axel Holst (1860–1931) and Theodor Frölich published a very famous paper concerning the use of guinea-pigs in the investigation of the causes of scurvy. Holst was professor of Hygiene and Bacteriology at the University of Christiana in Oslo. The choice of guinea-pigs for this study may look surprising today, when the ubiquitous rat is to be found in many biological laboratories. However at the end of the nineteenth century the rat was regarded as an unpleasant, dirty animal, whereas the guinea-pig was highly regarded as a childrens' pet and was cheap to buy and keep. The latter requirement was particularly important in the days before the existence of large research budgets which are available (at least in theory) today. The choice was of course extremely fortunate, since the guinea-pig is one of the few species of mammal which have lost, or never possessed, the ability to biosynthesise vitamin C. Holst and Fröhlich fed their guinea-pigs with the same diet which they had previously found had produced polyneuritis in pigeons. The symptoms of the disease induced in the guinea-pigs were very different

Harriette Chick (1875–1977)
Dr. N. Todhunter, Vanderbilt University)

Axel Holst (1860–1931)
(Dr. R. Forbes, University of Illinois)

Bottling lime juice in Victorian times
(Cambridge University Library)

from polyneuritis and included muscle haemorrhaging, brittle bones, and bleeding from the gums. Fröhlich had previously worked with children suffering from infantile scurvy and recognised the similarity in the pathological changes exhibited by guinea-pigs to those he had seen in children. This gave him and his co-workers the opportunity to test various diets to ascertain the extent of antiscorbutic activity. Thus they were able to demonstrate the effectiveness of fresh fruit and vegetables against scurvy conclusively. They showed that while fresh potatoes gave protection against scurvy, dried potatoes did not and that, in general, heating fresh fruit and vegetables effectively destroyed their antiscorbutic activity. The only conclusion which could be drawn from this work was that scurvy was produced by a defective diet.

Fredrick Gowland Hopkins was the first professor of biochemistry at the University of Cambridge. He studied carefully the influence of diets which are apparently well-balanced, containing purified protein, fat, carbohydrate, and minerals on the growth of rats. As we would now expect, the effect was catastrophic. However, the addition of only small amounts of cow's milk to the diet resulted in a complete recovery. It was becoming clear that there were a variety of substances which must be present in a healthy diet. It remained for Casimir Funk of the Lister Institute in London to propose in 1912 that diseases such as beri-beri, polyneuritis in birds, dropsy, scurvy, and pellagra are all deficiency diseases, which are prevented by the presence of small amounts of substances containing nitrogen in the diet. He proposed that all these compounds were, in fact, amines and that they be referred to as 'vital amines' abbreviated to 'vitamines'. Like many names which have stuck (dehydroascorbic acid is another), this turned out to be a misnomer and of course the 'e' was dropped and they are now called 'vitamins'.

Even at this late stage the list of mistaken causes of scurvy was being added to. McCollum, in 1917 proposed that scurvy was not a deficiency disease at all, but was due to constipation, which resulted in bacterial infection through the walls of the caecum, produced by the nature of the diet.

The First World War contributed to an acceleration in the understanding of the causes of scurvy in at least two ways. The first was that many researchers were to have first-hand experience of the havoc wrought by disease in many campaigns during the war. The second was that as men were recruited for the war effort abroad, women took on many of the tasks at home previously carried out by men. As a result of this a team of women remained at the Lister Institute in London, led by Harriette Chick, who developed an extensive research programme to investigate the dietary factors which were required to prevent the development of beri-beri and scurvy. Perhaps the most important feature of the experiments carried out at the Lister Institute at this time was that the workers took the greatest care to ensure that a very carefully balanced diet, containing no antiscorbutic foods, was fed to the guinea-pigs. This ensured that in all ways except contracting scurvy, the animals remained healthy. This allowed them to estimate the antiscorbutic *effectiveness* of foods and thus they were able to demonstrate that lemon juice was powerfully antiscorbutic, fresh potatoes moderately so, and fresh milk only slightly. They were able to settle once-and-for-all the question of the effectiveness of lime juice as an antiscorbutic. It turned out that West Indian sour limes were comparatively low in antiscorbutic power, and when the juice was preserved even this was destroyed. Hence, the problems experienced in many expeditions that took place at the end of the nineteenth and beginning of the twentieth centuries were due to the use of preserved West Indian lime juice. There was thus now a reliable biological assay for the antiscorbutic factor. It is no exaggeration to say that the careful application of this method of analysis provided information which almost immediately, when applied to food supplies to troops, saved many lives in some First World War campaigns.

The availability of a reliable biological method of assaying the antiscorbutic factor, although slow, was crucial in the eventual isolation of the vitamin. By 1919, McCollum had named two factors required in their diet for the survival of rats, 'A' and 'B'. Drummond, in that year suggested that the antiscorbutic factor be called 'water soluble C'. However, the term 'vitamin' for such essential factors was becoming widely adopted and the antiscorbutic factor became known as vitamin C.

The decade between 1920 and 1930 saw much activity in the investigation of vitamin C. Many attempts were made to isolate the vitamin. The Lister Institute was again at the forefront of the research and Zilva and his co-workers were able to explore aspects of the chemistry of the vitamin, without actually isolating it. They were able to show that it was a reducing agent and used this property, where appropriate, in analysis instead of the very time-consuming bioassay.

The isolation of the vitamin proved to be quite difficult and there were a number of 'near misses' among the attempts to prepare vitamin C. Perhaps the saddest of these was Link in the University of Wisconsin, who made crude calcium ascorbate, but was unable to show its antiscorbutic properties because the university refused him a grant for the bioassay of the substance. Another was when Vedder made a crude sample of vitamin C in 1927, but since he worked in the office of the Surgeon General of the United States Army, he was moved to another position before he was able to carry out the vital bioassays.

We come to the end of the decade tantalisingly close to the isolation of vitamin C. The next few years were to be filled with the drama of the isolation, structure determination, and synthesis of vitamin C, with controversy over who had been first to achieve the isolation.

Chapter 3

Discovery and Structure of Vitamin C

ISOLATION

The years immediately after the First World War saw research into the isolation of the elusive vitamin gaining momentum and, in both the USA and Europe, the race was really on to reap the tremendous scientific and financial rewards that would surely follow. We have seen that at the Lister Institute Solomon Zilva and his team were working feverishly to extract and purify the vitamin from concentrated citrus juices but, although they obtained material which was powerfully antiscorbutic, it was not a pure crystalline substance. Similar frustrations were being suffered by the leading American group, headed by Charles King at the University of Pittsburgh. One of the major problems was that, being a sugar-like substance, it was extraordinarily difficult to separate the vitamin from the many other different sugars present in concentrated fruit juices. Ironically, despite the dedicated and painstaking work of these groups, fate was setting the stage for an unknown Hungarian scientist, Albert Szent-Györgyi.

Born in 1893, into an aristocratic family of modest wealth, Albert Imre Szent-Györgyi von Nagyrapolt was a charismatic man who, after a distinctly unpromising school career in his native Budapest, had qualified as a doctor, his training having been rudely interrupted by some fairly hair-raising experiences as a soldier in the defeated Austro-Hungarian army during the First World War. Szent-Györgyi had that rare but essential gift of any great researcher, the ability, as he later put it, 'to see what everyone else has seen but think what no-one else has thought'.

After the war, he travelled across Europe with his wife and young daughter, working at various laboratories with some well-known scientists, and it was during this period that he developed a life-long

26

interest in the embryonic field of biochemistry and, in particular, the subject of intracellular respiration and the associated oxidation–reduction reactions occurring at the molecular level in living systems. Frequently short of money, but gaining valuable experience, he arrived in the Dutch University town of Groningen in 1922 where he acquired a post as assistant in physiology to Professor H. J. Hamberger. With his eye for detail, he perceived a possible, if unlikely, connection between the bronzing of patients suffering from Addison's disease (caused by defective adrenal gland function) and that of freshly cut potatoes, apples, and pears. It was known that this latter coloration arose from a disturbed redox process and he subsequently discovered, in the juices of lemons and oranges (which did not turn brown when damaged) and also in cabbage juice, a strong reducing agent. He then isolated the juices from the adrenal glands of cows and again found a reducing agent. He determined to isolate what he thought might be a new adrenal hormone.

After a brief and unsuccessful attempt to isolate his compound at Sir Henry Dale's London laboratories early in 1925, he returned to Groningen and, finding himself out of favour with the new professor, sent his family home to Budapest, resigned his position and went through a period of deep depression. However, a chance meeting at a conference in Stockholm with the now world-famous biochemist Professor Sir Frederick Gowland Hopkins (who had been impressed by one of his papers) resulted in an opportunity to work in Cambridge, whereupon Szent-Györgyi sent for his family and embarked on the work that would make his name.

After much frustration, Szent-Györgyi managed to accumulate less than a gram of an off-white crystalline substance from the adrenal cortex of cattle (in which it was present in only tiny amounts – about 300 mg per kilogram of starting material) and, later from orange juice and cabbage water. He described the process of extracting his 'reducing factor' as follows:

1. It was extracted from chilled and minced adrenal cortex by shaking with methyl alcohol, and bubbling through with carbon dioxide to prevent its having contact with oxygen.
2. The filtered extract was then mixed with a solution of lead acetate, which precipitated the reducing factor.
3. The precipitate, separated by further filtration, was suspended in water, and sulphuric acid was added. The reducing factor dissolved, and lead sulphate was precipitated.
4. The filtrate was evaporated to dryness in a vacuum.

5. The solids were re-extracted with methanol, and steps 2, 3, and 4 were repeated.
6. The solids were dissolved in acetone, and when an excess of light petroleum was added, crystals of the reducing factor gradually precipitated.

Like a typical reducing agent, the substance decolorised iodine and, from the combining masses of this reaction, he concluded that its relative molecular mass was 88.2 or a multiple thereof. The lowering of the vapour pressure of water by the crystals suggested a relative molecular mass of about 180, pointing to 176.4 as the correct value. Finally, combustion analysis, giving 40.7% carbon, 4.7% hydrogen, and 54.6% oxygen, enabled the molecular formula $C_6H_8O_6$ to be deduced.

Hopkins encouraged his protégé to publish this work describing what was then believed to be a new hormone with the nature of a sugar acid. Szent-Györgyi, however, incurred some difficulty getting his paper accepted because, in his first draft, he mischievously called his new compound 'ignose' (suggesting a sugar-like substance for which he didn't know the structure) and, when this was rejected, he resubmitted, naming the substance 'Godnose'! When the irritated editor finally threatened not to publish unless an appropriate name were chosen, Szent-Györgyi relented, accepting the editor's suggestion of 'hexuronic acid' and so, in the *Biochemical Journal* of 1928, this milestone of a paper appeared. No doubt this amusing anecdote featured prominently in many of Szent-Györgyi's future after-dinner speeches. It is interesting to note that, in this paper, Szent-Györgyi suggested that the reducing properties of fruit juices, of interest to 'students of vitamin C', could well be due to hexuronic acid. Two and two had not yet been put together but the idea that they were one and the same was surely germinating.

The following year, Szent-Györgyi visited the USA and spent some time at the Mayo Clinic in Rochester, Minnesota where, fortunately, the enormous nearby slaughterhouses furnished a plentiful supply of fresh adrenal glands. Szent-Györgyi was able to isolate almost 25 g of his hexuronic acid, untold riches, and he promptly sent about half back to England, to Professor Norman Haworth in Birmingham, for structure determination. Sadly, this quantity proved insufficient for Haworth's group to elucidate the structure, which remained a mystery.

Throughout his life, Szent-Györgyi seemed to have the good fortune of being in the right place at the right time and this was to prove so yet again. Following an approach from the Hungarian Minister of Education, Szent-Györgyi was presented with the chance to return home in triumph, in the summer of 1930, as Professor of Medical Chemistry at Szeged, about one hundred miles south of Budapest – an opportunity he readily accepted. He quickly gained a reputation as an unorthodox but approachable leader and an inspirational teacher, well-liked by both colleagues and students. Just over a year later, a former member of Professor King's group at Pittsburgh, Joe Svirbely, returning to study in his family's homeland, joined the group at Szeged. The thought that his ageing off-white crystals might be vitamin C had strengthened in Szent-Györgyi's mind but he was not particularly interested in vitamins and their associated media hyperbole and he certainly hated clinical trials. The arrival of Svirbely enabled this possibility to be rigorously tested and, by the spring of 1932, the true identity of 'hexuronic acid' had been established.

Out of loyalty, Svirbely felt obliged to inform King of his exciting news but there was horror and dismay at Szeged when King immediately published a letter in *Science* claiming the discovery that vitamin C and hexuronic acid were, in fact, the same compound. King, however, had written to Svirbely, just prior to receiving the latter's news, indicating that he was still uncertain as to whether vitamin C and hexuronic acid were identical and had some further work to do. This incident infuriated Szent-Györgyi and temporarily shattered his faith in the integrity and trustworthiness of the scientific community. However, the award of the 1937 Nobel Prize for Medicine to Szent-Györgyi 'in recognition of his discoveries concerning the biological oxidation processes with special reference to vitamin C' indicated where the scientific establishment as a whole believed the credit for discovery lay.

His supply of crystals now exhausted, Szent-Györgyi enjoyed another flash of uncanny intuition and associated good fortune. He discovered that Hungarian paprika, which grew in abundance locally, was particularly rich in the vitamin and, because of the lack of other sugars, its extraction and isolation from this source proved relatively easy. Within a week he had obtained well over a kilogram of pure crystals, virtually a world monopoly. Szent-Györgyi and Haworth renamed the substance 'ascorbic acid' (because it prevented scurvy) and, once more, Haworth's group set about determining the chemical structure, a quest which, unlike before, would soon prove successful.

STRUCTURE ELUCIDATION

The Birmingham group this time had in their possession an ample supply of Szent-Györgyi's off-white crystalline plates, a compound of known molecular formula ($C_6H_8O_6$), melting point (191 °C), and specific rotation ($+23°$ in water). Nevertheless, a fascinating piece of detective work, typical of the heroic age of organic chemistry, would be needed to unravel the structure. Edmund Hirst, who had worked under Haworth at Durham and subsequently rejoined him in Birmingham, was placed in charge of the work.

When boiled with hydrochloric acid, the crystals gave a quantitative yield of furfural showing that at least five of the six carbon atoms formed an unbranched chain. Further tests showed ascorbic acid to be a weak, monobasic acid and a strong reducing agent. The first stage of oxidation (which was easily reversible) could be brought about by, for example, aqueous iodine, acidified quinone, and molecular oxygen in the presence of copper salts at pH 5. The product of such oxidations, involving the loss of two hydrogens, was named dehydroascorbic acid ($C_6H_6O_6$). This reversible oxidation by iodine was analogous to a known reaction of 2,3-dihydroxymaleic acid (equation 1),

$$HOOCC(OH) = C(OH)COOH + I_2$$
$$\rightarrow HOOCCO\text{--}COCOOH + 2HI \qquad (1)$$

suggesting the possible presence of an ene-diol grouping, $C(OH)\text{-} = C(OH)$, and the similar absorption spectra of ascorbic and dihydroxymaleic acids, with a single, strong band at about 245 nm, reinforced this idea. Further supportive evidence for this grouping came from the rapid reactions with diazomethane, yielding dimethylascorbic acid, and with phenylhydrazine which, after first oxidising the ascorbic acid, reacted to give an osazone. These reactions are summarised in Figure 3.1 and expanded upon in Chapter 4.

The acidity of ascorbic acid initially suggested the presence of a carboxylic acid group but dehydroascorbic acid was shown to be a neutral lactone which actually hydrolysed slowly to a carboxylic acid. The easy, smooth interconversion of ascorbic and dehydroascorbic acids strongly pointed to the former also being a lactone, a view supported by the fact that dimethylascorbic acid was a neutral compound which reacted with sodium hydroxide to give a sodium salt without loss of a methyl group, indicating lactone ring-opening (Figure 3.2).

Sir Norman Haworth (1883–1950)
(Wellcome Institute Library, London)

Albert Szent-Györgyi (1893–1986)
(Wellcome Institute Library, London)

Sir Edmund Hirst (1898–1975)
(Maurice Stacey, personal collection)

Nobel Prizewinners 1937: seated, from left are Roger Martin du Gard (Literature),
Albert Szent-Györgyi (Physiology and Medicine), Paul Karrer and Norman
Haworth (Chemistry), and Clinton J. Davisson (Physics)
(National Library of Medicine)

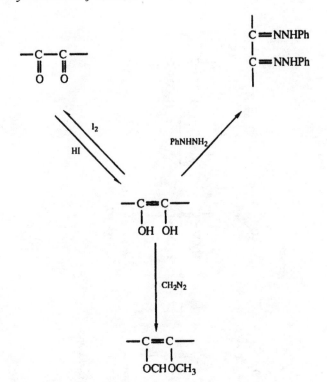

Figure 3.1 *Reactions of the ene-diol group*

It was also known that there were two more alcoholic OH groups present, these reacting with acetone (propanone) to give an isopropylidene derivative which still contained the two enolic hydroxyls.

Further oxidation of dehydroascorbic acid, using sodium hypoiodite in alkaline solution, yielded oxalic acid and L-threonic acid, the latter being identified by its conversion into the known substances L-dimethoxysuccinamide and tri-*O*-methyl-L-threonamide. This established the stereochemical relationship of natural ascorbic acid to the L-series of sugars and also confirmed that the lactone carbonyl was directly adjacent to the ene-diol grouping (Figure 3.3).

Evidence was now needed for the size of the lactone ring and this emerged as a result of another elegant piece of organic chemistry. It was known that diazomethane converted ascorbic acid into a di-*O*-methyl derivative and further methylation with iodomethane and silver oxide produced a tetra-*O*-methylated compound which, on

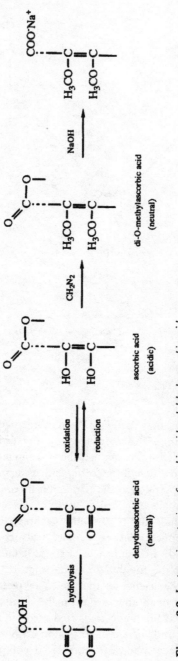

Figure 3.2 *Lactone ring-opening of ascorbic acid and dehydroascorbic acid*

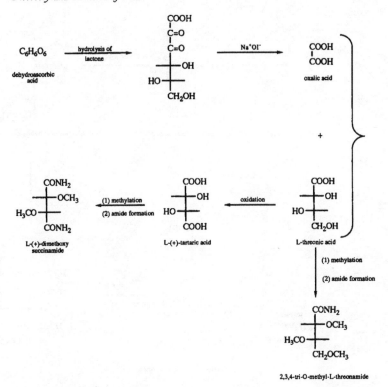

Figure 3.3 *Cleavage of dehydroascorbic acid and establishment of ascorbic acid configuration*

ozonolysis, gave a *single*, neutral ester. Degradation of this product with methanolic ammonia yielded oxamide and 3,4-di-O-methyl-L-threonamide, the latter being identified by the fact that it gave the Weerman reaction, characteristic of 2-hydroxyamides. This confirmed the point of attachment of the lactone ring as being at C-2 of the threonamide, equivalent to C-4 of the tetra-O-methylascorbic acid (Figure 3.4).

Ascorbic acid was, therefore, shown to be a γ-lactone, conveniently represented as in Figure 3.5 although other tautomeric forms may exist in small quantities. The configuration at C-5 is L- (or S-, using the Cahn–Ingold–Prelog system). The acidic nature in aqueous solution derives from the ionisation of the enolic OH on C-3 (pK_a 4.25), the

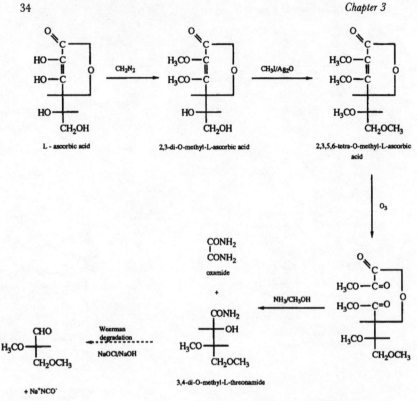

Figure 3.4 *Establishment of lactone ring size of ascorbic acid*

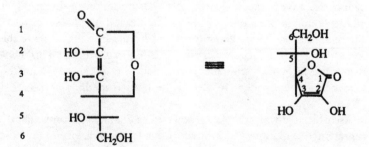

Figure 3.5 *Structure of L-ascorbic acid*

Table 3.1 *Physical data for L-ascorbic acid*

Nomenclature	Hexuronic acid, cevitamic acid, redoxon; L-ascorbic acid, L-*xylo*-ascorbic acid, L-*threo*-2,3,4,5,6-pentahydroxyhex-2-enoic acid-4-lactone L-*threo*-hexono-1,4-lactono-2-ene
m.pt. (°C)	190–192 (with decomposition)
rel.mol.mass	176.14
pK_a	1st (C-3–OH) 4.25
	2nd (C-2–OH) 11.79
Specific rotation	$[\alpha]_D^{20} = +23°$ in water; $[\alpha]_D^{18} = +49°$ in methanol
Solubility	33 (water), 3 (ethanol), 1 (glycerol)
(g per 100 cm³ at 20°C)	insoluble in chloroform, benzene, ether, petroleum ether, fats and oils
Density	1.65 g cm⁻³

resulting ascorbate anion being delocalised (Figure 3.6). A summary of the physical properties of L-ascorbic acid is given in Table 3.1.

The Birmingham group quickly confirmed this structure by developing a synthetic route to L-ascorbic acid. This work, published in 1933, was the end-product of much painstaking and laborious synthetic chemistry and it was therefore singularly appropriate that Norman Haworth should find himself sharing the same stage as Szent-Györgyi in Stockholm in 1937, receiving the Nobel Prize for Chemistry for his work on 'the structure of carbohydrates and vitamin C'. Albert Szent-Györgyi himself was to make several more major scientific breakthroughs, doing most of the spadework leading to the discovery of the Krebs cycle, pioneering the biochemistry of muscle, and carrying out much innovative, basic research into the causes of, and possible cures for, cancer.

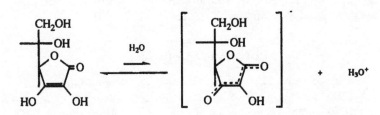

Figure 3.6 *First ionisation of L-ascorbic acid*

After the Second World War, he settled in the USA and became an American citizen. This most remarkable man, genuinely one of the giants of twentieth-century science with a uniquely natural gift for the intuitive hunch, was still working on a new paper shortly before his death in 1986 at the age of ninety-three.

STRUCTURAL DETAIL BY INSTRUMENTAL METHODS

The growth and development of spectroscopic techniques in the years following the successful elucidation of the L-ascorbic acid structure have enabled that structure to be examined and understood more deeply.

X-ray diffraction

The first *X*-ray analyses of L-ascorbic acid were carried out in the early 1930s in an attempt to assist the structural investigations being pursued in Birmingham and did, indeed, support the conclusions of this group. More recent *X*-ray and neutron diffraction studies by Hvoslef in Oslo in the 1960s have revealed a monoclinic crystal structure with four molecules per unit cell. There are two types of molecule (A and B) in the crystal with eight distinct intermolecular hydrogen-bonds (Figure 3.7). The conformations of the A and B molecules are virtually identical with the C-5–OH anti-periplanar with respect to both C-4–H and C-6–OH (Figure 3.8).

Salts of L-ascorbic acid contain the resonance-stabilised ascorbate anion formed on ionisation of the C-3–OH proton. *X*-Ray data reveal the anticipated lengthening and shortening of the bonds along the conjugated O–C-3$=$C-2–C-1$=$O system of the anion as compared with the neutral molecule (Table 3.2).

In many ascorbate salts, the combined effects of metal co-ordination and hydrogen-bonding apparently cause the C-6–OH to adopt a *syn-clinal* (or gauche) position with respect to the C-5–OH.

Ultraviolet Spectroscopy

The u.v. spectrum of L-ascorbic acid at pH 2.0 reveals a λ_{max} of 243 nm ($\epsilon = 10\,000$ mol^{-1} dm^3 cm^{-1}) which undergoes a red shift to 265 nm ($\epsilon = 16\,500$ mol^{-1} dm^3 cm^{-1}) at pH 7.0 as a result of ionisation of the C-3–OH proton.

These figures are consistent with $\pi \rightarrow \pi^*$ excitation of a conjugated

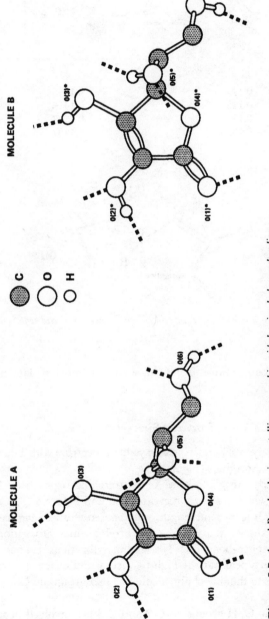

Figure 3.7 *A and B molecules in crystalline L-ascorbic acid showing hydrogen-bonding* (After Falk and Wójcik, *Spectrochim. Acta*, 1979, **35A**, Pergamon Press Plc)

Table 3.2 *Bond lengths* (nm) *in* L-*ascorbic acid and sodium ascorbate*

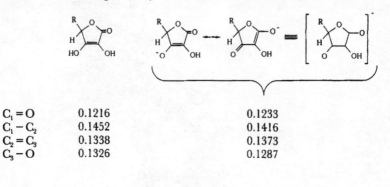

$C_1 = O$	0.1216	0.1233
$C_1 - C_2$	0.1452	0.1416
$C_2 = C_3$	0.1338	0.1373
$C_3 - O$	0.1326	0.1287

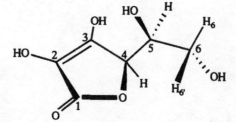

Figure 3.8 *Structure of* L-*ascorbic acid molecule showing preferred conformations*

carbon–carbon double bond in a five-membered lactone ring structure.

Infrared Spectroscopy

Infrared analysis of L-ascorbic acid yields a spectrum with a number of interesting absorptions (Figure 3.9).

Particularly intriguing is the O–H stretching region from 4000 to 2000 cm^{-1} which is shown in greater detail in Figure 3.10.

As the hydrogen-bond lengths (and consequently strengths) in the crystal are known, it is possible to correlate these absorptions with specific hydrogen-bonded O–H stretching vibrations. It appears that the four sharp peaks at the high-frequency end of the spectrum are attributable to the side-chain, alcoholic O–H groups on C-5 and C-6 (Table 3.3).

The enolic O–H groups on C-2 and C-3 are involved in stronger

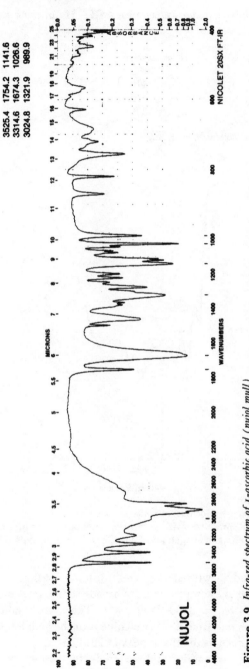

Figure 3.9 *Infra-red spectrum of L-ascorbic acid (nujol mull)*
(Reproduced with permission, from the Aldrich Chemical Company Inc.)

Table 3.3 *High-frequency hydrogen-bonded* O–H *stretching bands of* L-
ascorbic acid

Bond	Wavenumber/cm⁻¹	O–O distance/nm
O-6–H···O	3525	0.294
O-5–H···O	3410	0.279
O-6*–H···O	3315	0.277
O-5*–H···O	3220	0.271

Asterisks indicate molecule B in (Figure 3.7).

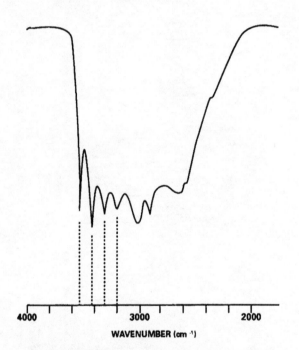

Figure 3.10 *Infra-red spectrum of* L-*ascorbic acid* (NaCl *disc*)
(Reproduced with permission, in modified form, from
Lehmann, Pagel, and Penka, *Eur.J.Biochem*, 1984, **138**, 480)

hydrogen-bonding with shorter O–O distances of 0.261–0.267 nm.
These give rise to a complex series of broader bands from molecules A
and B in the region 3100–2200 cm⁻¹. There are doubtless some
overlapping signals from C–H stretching near the high frequency end
of this region such as the sharp peak at 2915 cm⁻¹.

Turning to lower-wavenumber signals, the strong absorption at

1754 cm⁻¹ has been attributed to $C=O$ stretching of the five-membered lactone ring system with the intense doublet at 1675 and 1660 cm⁻¹ arising from $C=C$ stretching vibration (coupled with neighbouring vibrations along the conjugated system). The medium signal at 1460 cm⁻¹ is attributed to CH_2 scissoring.

The fingerprint region is complex but attempts have been made to correlate specific vibrations, *e.g.* 1320 cm⁻¹ (O–H deformation of C-2–OH), 1275 cm⁻¹ (C-2–O stretch), 1140 cm⁻¹ (C-5–O stretch), and 1025/990 cm⁻¹ (lactone ring deformation).

Nuclear Magnetic Resonance Spectroscopy

The proton-decoupled ¹³C n.m.r. spectrum of L-ascorbic acid shows six signals as expected and chemical shifts have been assigned (Figure 3.11).

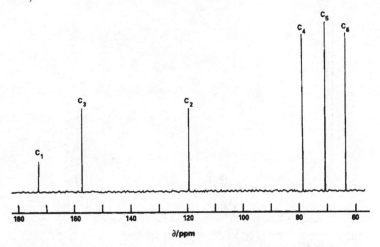

Figure 3.11 *¹³C n.m.r. spectrum of aqueous L-ascorbic acid (proton decoupled) at pH 2*

Off-resonance decoupling produces the anticipated splitting pattern *i.e.* C-1, C-2, and C-3 (singlets), C-4 and C-5 (doublets), and C-6 (triplet). C-4 and C-5 were distinguished by preparing the 4-D-derivative and again recording the proton-decoupled spectrum whereby the signal at 77δ p.p.m. became a triplet due to deuterium coupling. C-3 was identified by the large (19δ p.p.m.) downfield shift of its signal observed when the pH is changed from 2 to 7 causing ionisation of the C-3–OH proton.

The ¹H n.m.r spectrum of L-ascorbic acid is particularly interesting and careful analysis has enabled the conformations of the molecule in aqueous solution to be studied. When the spectrum is run in D_2O, the four OH protons are exchanged and do not appear as separate signals. The remaining four protons (H-6, H-6', H-5, and H-4) comprise an ABMX system, the two protons on C-6 being non-equivalent because of the chirality of C-5.

The fine structure of these signals is not observable at low field strengths (60 or 100 MHz) but above 300 MHz the non-first-order, spin-spin pattern is revealed. This pattern is particularly clearly seen in the ¹H n.m.r spectrum of sodium ascorbate (Figure 3.12).

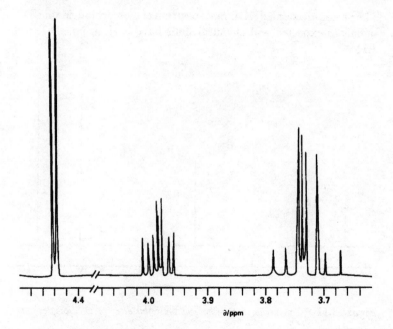

Figure 3.12 ¹H *n.m.r. spectrum of sodium ascorbate* (D_2O)
                 (Reproduced with permission from Reid, *J.Chem.Ed.*, 1989,
                 66, 345)

The magnitude of the proton coupling constants, *J*, obtained from such spectra enables the preferred conformation of L-ascorbic acid in aqueous solution to be deduced. For example, $J_{H4, H5}$ is found to be 1.8 Hz which is the predicted value for the conformation shown in

Table 3.4 *Correlation of coupling constants with* C-5–C-6 *conformation*

	$J_{H5, H6}$/Hz	$J_{H5, H6'}$/Hz	Population of conformer %
Conformer I (theoretical)	5.0	10.8	58
Conformer II (theoretical)	10.8	3.0	25
Conformer III (theoretical)	0.8	2.9	17
observed values (^{1}H n.m.r.)	5.7	7.5	

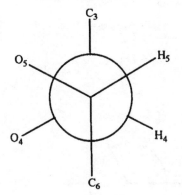

Figure 3.13 *Preferred* C-4–C-5 *conformations of aqueous* L-*ascorbic acid*

Figure 3.13. Thus the preferred conformation about the C-4–C-5 bond in aqueous solution is the same as that observed in the crystal (Figure 3.8).

Equally informative is the correlation of coupling constants with conformation about the C-5–C-6 bond. The possible stable conformations are shown (Figure 3.14). Assuming that the observed coupling constants represent a weighted average of the theoretical values for the three conformations, the populations of these conformations can be estimated (Table 3.4). Once again the preferred conformation about the C-5–C-6 bond in solution is identical with that found in the crystal.

The similarity between the preferred conformations of the side-chain in both crystal lattice and aqueous solution is probably a consequence of the lack of intramolecular hydrogen-bonding in both environments. Considerable intermolecular hydrogen-bonding occurs, of course, with neighbouring ascorbic acid molecules in the crystal and with water molecules in aqueous solution.

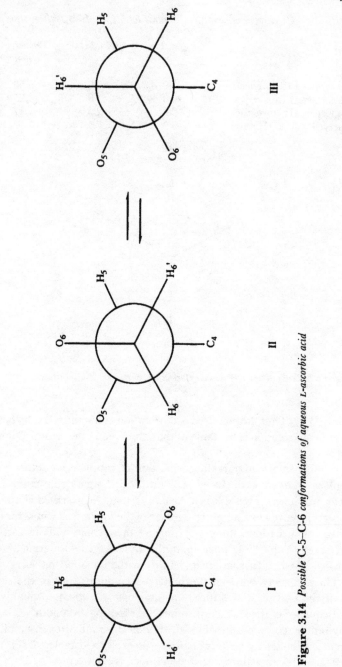

Figure 3.14 *Possible C-5–C-6 conformations of aqueous L-ascorbic acid*

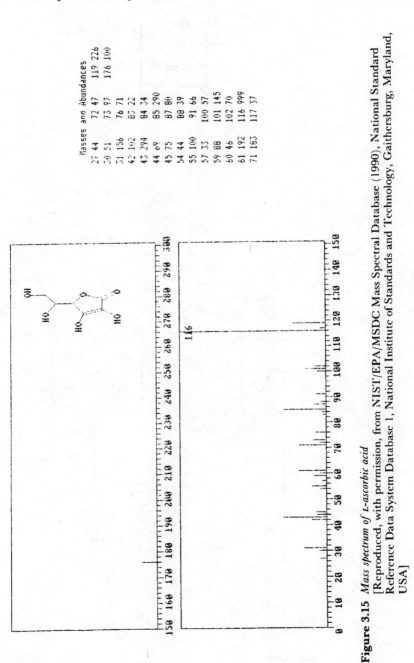

Masses	and	Abundances			
27	44	72	47	119	226
50	51	73	97	176	100
51	156	76	71		
42	192	83	22		
43	294	84	34		
44	69	85	290		
45	75	87	80		
54	44	88	39		
55	100	91	66		
57	33	100	57		
59	88	101	145		
60	46	102	70		
61	192	116	999		
71	183	117	37		

Figure 3.15 *Mass spectrum of L-ascorbic acid*
[Reproduced, with permission, from NIST/EPA/MSDC Mass Spectral Database (1990), National Standard Reference Data System Database 1, National Institute of Standards and Technology, Gaithersburg, Maryland, USA]

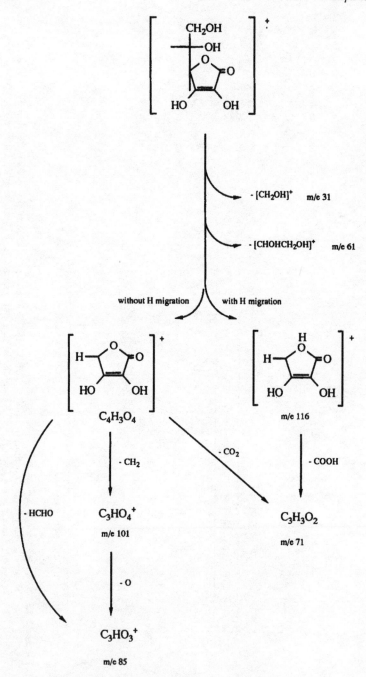

Figure 3.16 *Possible mass spectral fragmentation of L-ascorbic acid*

Mass Spectrometry

The electron-impact mass spectrum of L-ascorbic acid shows a clear molecular ion at m/e 176 together with a base-peak at m/e 116 (Figure 3.15). Fragmentation involving loss of the side chain followed by ring rupture has been suggested and is consistent with a number of the more significant signals (Figure 3.16).

Chapter 4

Synthesis, Manufacture, and Further Chemistry of Vitamin C

SYNTHETIC METHODS FOR L-ASCORBIC ACID

In 1933, the Swiss group of Tadeus Reichstein reported the synthesis, from D-xylosone, of a substance that was identical with natural ascorbic acid except for its specific rotation. The precise chemical structure of ascorbic acid was not known to them at the time (in fact they ascribed an incorrect structure to their product) and it soon became clear that they had actually synthesised the D-isomer of the vitamin. Later that same year, Norman Haworth and Edmund Hirst, a formidable research partnership, supported by their large team of skilled synthetic chemists at the Frankland Laboratories in Birmingham, and armed with their knowledge of the true structure of ascorbic acid, announced in *Chemistry and Industry* that they had accomplished the synthesis of the natural substance. They used L-xylosone as their starting material, painstakingly prepared from D-galactose via a tortuous synthetic route involving eight steps (Figure 4.1), and then utilised the same three-stage reaction sequence as the Swiss chemists (Figure 4.2).

The leadership of this synthesis team was assigned to Maurice Stacey, a young, post-doctoral chemist who had, in fact, started life at Birmingham as an undergraduate. Haworth drove the team mercilessly hard, night and day, and they endured not only exhaustion from the work itself but also the discomfort of skin lesions and boils caused by persistent exposure to phenylhydrazine. Despite having practised with the D-isomer, considerable difficulty was experienced in getting the eagerly sought L-product to crystallise but, after much frustration, Stacey was at last able to inform Haworth that he had obtained a small quantity of the elusive crystals, identical in every respect with natural L-ascorbic acid. The work was published in haste as the Birmingham

48

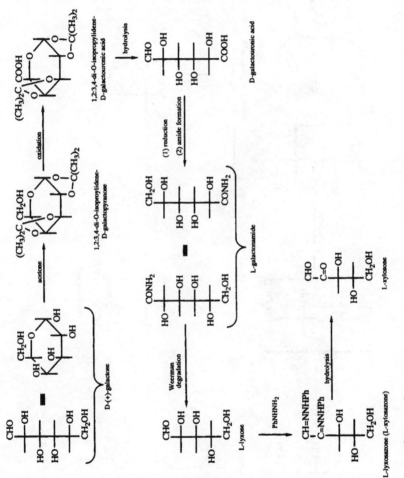

Figure 4.1 *L-Xylosone from D-galactose*

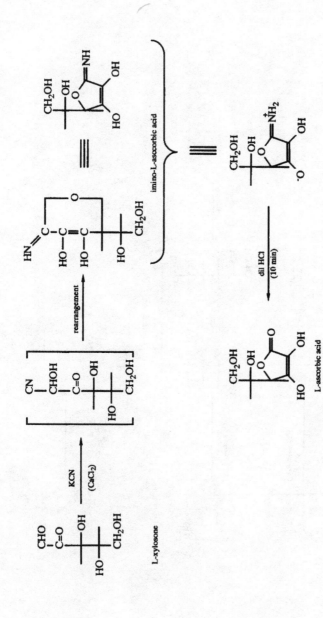

Figure 4.2 *L-Ascorbic acid from L-xylosone*

group knew that rival groups in Switzerland and Germany were within an ace of emulating their achievement. In his paper, Haworth acknowledged, in addition to himself and Stacey, the dedicated work of no fewer than seven chemists – R. G. Ault, D. K. Baird, H. C. Carrington, R. Herbert, E. L. Hirst, E. G. V. Percival, and F. Smith.

In later years, both Haworth and Hirst received knighthoods and several members of the team progressed to University Chairs, including Stacey himself who was to become Mason Professor of Chemistry at Birmingham from 1956 to 1974. Although the Frankland building no longer houses the chemical laboratories at Birmingham, the marvellous work of Haworth's School of Carbohydrate Chemistry is honoured by a commemorative plaque, most appropriately unveiled in 1981, seven years after his retirement, by Professor Maurice Stacey, CBE, FRS.

The synthesis of ascorbic acid from C_5 sugars is now well established, xylose, lyxose, and, recently (1980), arabinose being used as precursors.

About a year after Haworth and Hirst's success, Reichstein and Grüssner described a new route to L-ascorbic acid from the C_6 sugar D-(+)-glucose (Figure 4.3). This remains the most important synthetic method yet devised for vitamin C and, with some modifications, still provides the basis of the modern process for its commercial manufacture. Indeed, Reichstein, in collaboration with the pharmaceutical giant Hoffman-La Roche, subsequently filed a series of patents for the production of vitamin C which ultimately proved to be very lucrative. The process involved the reduction of D-glucose to D-glucitol (D-sorbitol) which was then fermentatively oxidised to L-sorbose with *Acetobacter suboxydans* (or *Acetobacter xylinum*). Di-O-isopropylidenyl protection of the hydroxy-groups on carbons 2,3 and 4,6 allowed the smooth oxidation of the C_1 primary alcohol to be carried out with potassium permanganate. Deprotection, esterification, and enolisation (with sodium methoxide) followed by acidification gave ascorbic acid in 15–18% yield (Figure 4.3).

In the modern industrial manufacturing process, the reduction to D-sorbitol is accomplished either electrolytically or by catalytic hydrogenation $(H_2/CuCrO_2)$. Additionally, it has been found that the reaction of sorbose with acidified acetone at low temperature $(-5\,°C)$ gives a greatly increased yield of the di-O-isopropylidenyl (as opposed to the 2,3-mono-O-isopropylidenyl) derivative. The oxidation of this protected sorbose (2,3:4,6-di-O-isopropylidene-L-sorbofuranose or 'diacetone sorbose') to the corresponding 2,3:4,6-di-O-isopropylidene-2-

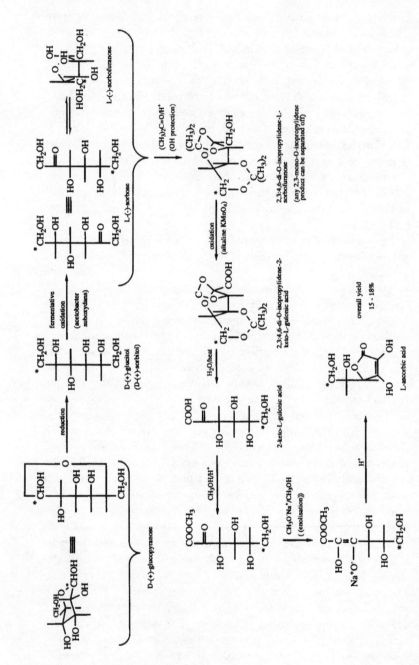

Figure 4.3 *L-Ascorbic acid from D-glucose with carbon-chain inversion*

Tadeus Reichstein
(Urs Schachenmann, F. Hoffmann-
La Roche AG, Basle)

Maurice Stacey CBE, FRS
(Maurice Stacey, personal collection)

Roche, Dalry: the buildings used for vitamin C synthesis run from left to right at the top of the photograph (Urs Schachenmann, F. Hoffmann-La Roche AG, Basle)

keto-L-gulonic acid (or 'diacetone-keto-gulonic acid') is commonly accomplished using bleach (sodium hypochlorite), produced on-site, in the presence of a nickel sulphate catalyst. It has also now been found that this protected acid can be smoothly converted directly into ascorbic acid using effectively non-aqueous conditions (*e.g.* chloroform–ethanol, concentrated hydrochloric acid, 65 °C, 50 hours). The crude acid is then isolated, purified by decolorisation with activated carbon, and, finally, recrystallised from water. As a result of these modifications, each step in the manufacturing process (outlined in Figure 4.4) now gives over 90% yield, resulting in greater than 50% conversion of the original glucose.

It has been estimated that about forty thousand tonnes of the vitamin are now manufactured annually (not including any produced in China or those countries that, until recently, were regarded as the 'Eastern bloc') and one of the biggest producers, not unexpectedly, is the Roche Company. Their vitamin C plant at Dalry, in the pleasant, rural setting of North Ayrshire in Scotland, came on-stream in 1983 as the third phase of a major development which already produced vitamins A, B_1, and B_5 and it is now the largest single producer of vitamin C in the Roche organisation. More than 90% of Dalry's output is exported from the UK for use in food, animal feedstuffs, and various vitamin C and multivitamin preparations.

A careful consideration of the Reichstein synthesis shows that C-1 of the D-glucose precursor molecule becomes C-6 in the L-ascorbic acid product. This is known as carbon-chain inversion and a number of further synthetic procedures have been reported in recent years involving similar glucose C-1/C-6 inversion. The path of the glucose C-1 carbon is asterisked in Figure 4.3.

In addition to these C-1/C-6 inversion syntheses from D-glucose, several synthetic routes, also from glucose, have been devised which do not involve carbon-chain inversion. These conversions, in which the C-1 of glucose remains the C-1 of ascorbic acid, basically involve the oxidation of glucose at C-1 and C-2 and inversion of configuration at C-5, the latter being accomplished by oxidation followed by stereoselective reduction (Figure 4.5). Although such syntheses have not proved to be of commercial value, they have found use in the preparation of C-5 deuterated derivatives of L-ascorbic acid.

Other than D-glucose, only one C_6 precursor, D-galactouronic acid [CHO(CHOH)$_4$COOH], obtained by enzymic hydrolysis of pectin, has been usefully converted into L-ascorbic acid but the overall yield is poor.

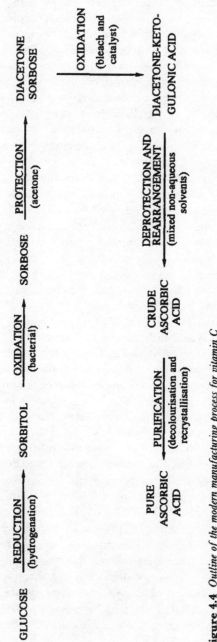

Figure 4.4 *Outline of the modern manufacturing process for vitamin C*

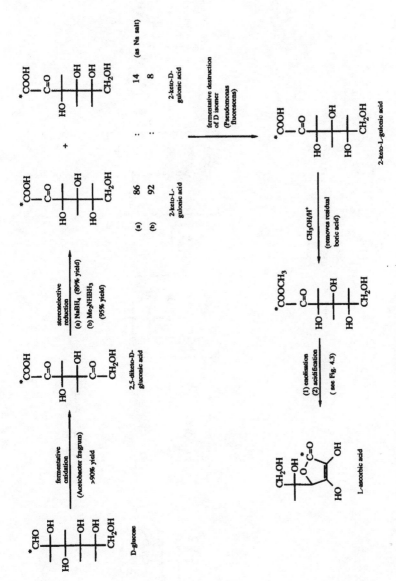

Figure 4.5 *L-Ascorbic acid from D-glucose without carbon-chain inversion*

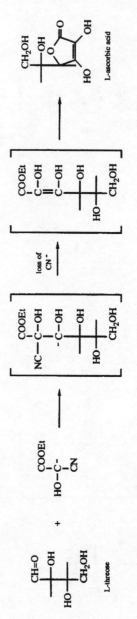

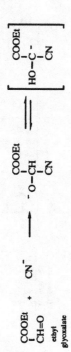

Figure 4.6 *L-Ascorbic acid from C4 and C2 precursors*

A third basic synthetic approach to L-ascorbic acid involves the combination of C_4 and C_2 carbon units. An example of this is the benzoin condensation reaction between L-threose and ethyl glyoxalate in the presence of sodium cyanide (Figure 4.6).

Interestingly, although quite a number of different synthetic procedures to L-ascorbic acid have been described since Haworth and Hirst first successfully achieved the task over fifty years ago, they all basically fall into three categories involving either the direct conversion of a C_6 precursor or the combination of appropriate C_1 and C_5 or C_2 and C_4 units (Figure 4.7).

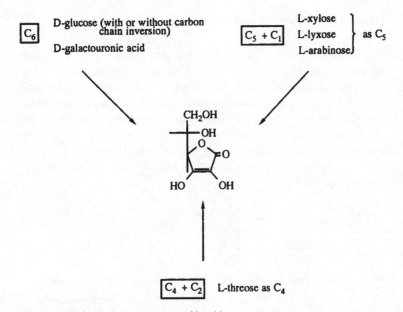

Figure 4.7 *Synthetic routes to L-ascorbic acid*

FURTHER CHEMISTRY OF L-ASCORBIC ACID

Some reactions of L-ascorbic acid have been mentioned in connection with the original structure determination of the molecule. These and others will be expanded upon in the following sections.

Alkylation and Acylation

As with many carbohydrates, the primary alcohol group at C-6 undergoes ready triphenylmethylation (tritylation) with chlorotriphenylmethane and pyridine (Figure 4.8).

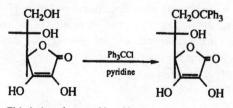

Figure 4.8 *Tritylation of L-ascorbic acid*

Methylation of L-ascorbic acid with diazomethane has thrown some light on the tautomeric nature of the vitamin. The greater acidity of the C-3–OH enables it to be titrated with ethereal diazomethane to give 3-*O*-methylascorbic acid. This reaction, however, is also accompanied by the formation of a small quantity of 1-methyl-ψ-L-ascorbic acid, presumably from a minor tautomer. Both of these compounds undergo further methylation with ethereal diazomethane to give 2,3-di-*O*-methyl-L-ascorbic acid and 1,2-di-*O*-methyl-ψ-L-ascorbic acid respectively (Figure 4.9).

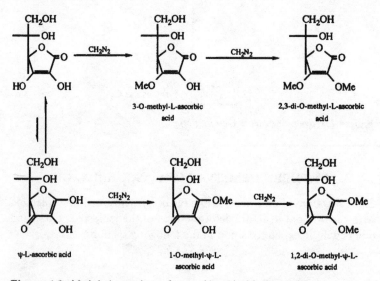

Figure 4.9 *Methylation products of L-ascorbic acid with diazomethane*

The 2,3-di-*O*-methyl derivative undergoes an interesting sequence of reactions on treatment with alkali and subsequent acidification. Rather than reforming the simple monocyclic lactone, a bicyclic derivative (with only one free hydroxyl), 2,3-isodimethyl-L-ascorbic acid, is produced. Acid hydrolysis of this product gives 2-*O*-methyl-L-ascorbic acid which is also obtained when cold, aqueous 1,2-di-*O*-methyl-ψ-L-ascorbic acid loses its labile methyl on C-1 on standing. As expected, 2-*O*-methyl-L-ascorbic acid readily gives the 2,3-di-*O*-methylated derivative with diazomethane in ether. These changes are summarised in Figure 4.10.

2,3-di-*O*-methyl-L-ascorbic acid undergoes both further methylation with iodomethane and silver oxide to give the 2,3,5,6-tetra-*O*-

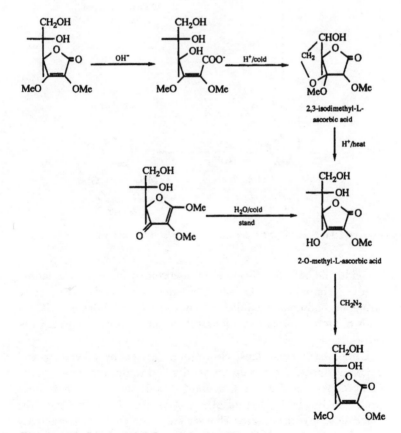

Figure 4.10 *Reactions of di-O-methyl-L-ascorbic acids*

methylated product and tritylation of the primary hydroxy-group on C-6. Methylation followed by detritylation with acid provides a route to the 2,3,5-tri-*O*-methyl derivative (Figure 4.11). This 2,3,5-tri-*O*-methyl compound has been shown to give a bicyclic 2,3,5-isotrimethyl derivative as part of a sequence of reactions identical to that outlined in Figure 4.10.

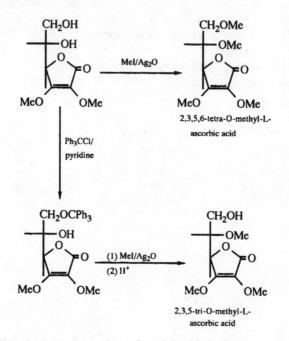

Figure 4.11 *Synthesis of tri- and tetra-O-methylated derivatives*

Acid-catalysed esterification of ascorbic acid, *e.g.* acetylation, produces initially the O-6-acylated derivative and, under more stringent conditions, a 5,6-diester. The crystalline 5,6-diacetate is well known. More vigorous conditions still are needed to obtain the 2,3,4,6-tetra-acetate.

Under basic conditions, electrophilic attack by alkylating and acylating agents depends on the acidity and steric accessibility of the OH groups in the C-2, C-3, C-5, and C-6 positions. Although the most acidic hydrogen is that of the C-3 hydroxyl ($pK_a = 4.25$), the delocalisation of the negative charge in this anion reduces its reactivity and also introduces an ambident nature which can lead for example to both C-2 and O-3 alkylation. As a result, reaction at the ascorbate O-3 only occurs with very reactive alkylating and acylating agents (such as diazomethane and acid chlorides) and selective O-3 derivatisation is generally difficult.

Removal of a second proton from the C-2 hydroxyl ($pK_a = 11.79$) produces a dianion which reacts preferentially at the more reactive oxygen on C-2 allowing selective alkylation/acylation at this position (Figure 4.12). A similar approach can be employed to synthesise selectively certain 2-O-inorganic esters such as the 2-O-sulphate. If both the C-2 and C-3 hydroxy-groups are protected, then base-promoted alkylation or acylation takes place at the more sterically accessible primary hydroxyl on C-6 rather than C-5, reaction at this latter position occurring only after derivatisation of the other three (Figure 4.13).

Acetal and Ketal Formation

The acid-catalysed formation of acetals and ketals of ascorbic acid is particularly useful for the selective protection of the molecule whilst structural modification is being carried out. The 5,6-O-derivatives such as the isopropylidene ketal and benzylidene acetal are well known but, more recently, it has proved possible to protect the C-2/C-3 positions selectively using particularly reactive aldehydes (Figure 4.14). This new development has paved the way for selective modification of the primary and secondary alcohol groupings on the ascorbic acid side chain.

Oxidation

L-Ascorbic acid is a powerful reducing agent in aqueous solution, this property being much less evident in non-aqueous media. The first stage of its oxidation is readily reversible and yields dehydroascorbic acid whose structure and properties are discussed in a later section. This first oxidation product still possesses reducing ability, especially in alkaline solution, being degradatively oxidised by, for example, molecular oxygen and hypoiodite ion to L-threonic and oxalic acids – a fragmentation which proved significant in the original structure determination of vitamin C. Alkaline hydrogen peroxide and acidic or alkaline permanganate also bring about oxidative cleavage, a number of different oxidation products being detectable in addition to the above mentioned acids.

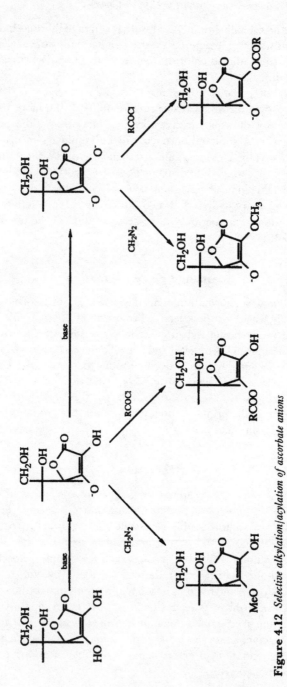

Figure 4.12 *Selective alkylation/acylation of ascorbate anions*

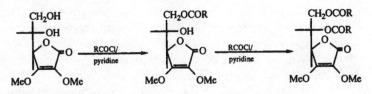

Figure 4.13 *Acylation at C-5 and C-6*

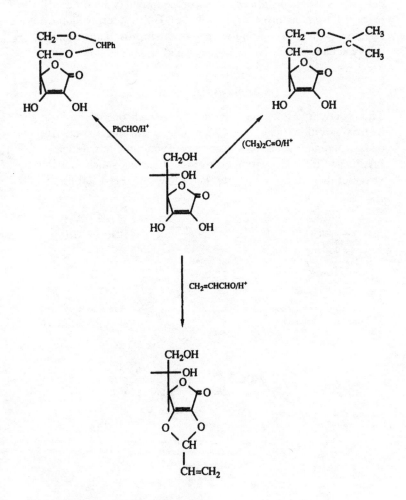

Figure 4.14 *Acetal and ketal formation*

The rate of aerobic oxidation of ascorbic acid depends on the pH, exhibiting maxima at pH 5 and 11.5. However, the reaction is much more rapid and the degradation more extensive in alkali. Degradative oxidation also occurs, albeit slowly, under anaerobic conditions.

Irradiation of aqueous ascorbic acid with ultraviolet, *X*-, and *γ*-radiation has been found to cause free-radical, photochemical oxidation under both aerobic and anaerobic conditions.

Oxidation of the primary and secondary alcohol groups on C-5 and C-6 could, theoretically, lead to a number of side-chain oxidised derivatives of L-ascorbic acid. Such compounds may not necessarily be prepared from ascorbic acid itself but C-2/C-3 protection would certainly be necessary if that were to be attempted. Some examples of this type of compound have been isolated (Figure 4.15).

Deoxy Compounds

Those derivatives of L-ascorbic acid in which one or more of the hydroxy-groups on C-2, C-3, C-5, or C-6 are missing are described as deoxy compounds. The lack of stability of vitamin C has, understandably, fuelled interest in analogues of the molecule which, possibly by virtue of their deoxy structures, may show enhanced stability whilst

6-carboxy-L-ascorbic acid
(saccharoascorbic acid)

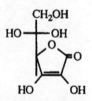

5-keto-L-ascorbic acid
(hydrated form isolated)

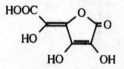

6-carboxy-5-keto-L-ascorbic acid
(5-ketosaccharoascorbic acid)
(enol tautomer stable)

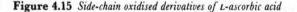

Figure 4.15 *Side-chain oxidised derivatives of L-ascorbic acid*

still retaining antiscorbutic activity. It has been known for some time that both 6-deoxy-L-ascorbic acid and L-rhamnoascorbic acid (Figure 4.16) possess about 0.3 and 0.2 times the antiscorbutic activity of L-ascorbic acid respectively. Examples of such derivatives that have been isolated more recently, and their interconversions are illustrated (Figure 4.17). Encouragingly, the 6-chloro-6-deoxy derivative has already been found to have a greater thermal stability than vitamin C whilst still retaining marked antiscorbutic properties.

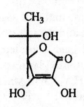
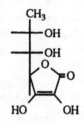

6-deoxy-L-ascorbic acid

L-rhamnoascorbic acid
(a 7-deoxy compound)

Figure 4.16 *Some well-established antiscorbutic deoxy derivatives*

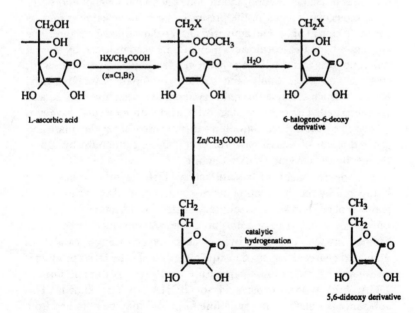

L-ascorbic acid

HX/CH₃COOH
(x=Cl,Br)

H₂O

6-halogeno-6-deoxy
derivative

Zn/CH₃COOH

catalytic
hydrogenation

5,6-dideoxy derivative

Figure 4.17 *Recent syntheses of antiscorbutic deoxy derivatives*

DEHYDROASCORBIC ACID AND ITS DERIVATIVES

Preparation

During his work at the Lister Institute in the mid-1920s, Solomon Zilva had observed that oxidised solutions of vitamin C retained their antiscorbutic activity. At about the same time, Szent-Györgyi reported that his 'hexuronic acid' could be oxidised by iodine with the loss of two hydrogen atoms and that the product could be quantitatively reduced by hydrogen sulphide back to starting material. Once the fact that vitamin C and hexuronic acid were one and the same became established, and the structure elucidation had been completed, the nature of this redox equilibrium became clearer, the oxidation product being identified as L-dehydroascorbic acid (or DHA). DHA was shown to be a lactone which underwent slow hydrolysis to a carboxylic acid on standing in aqueous solution and so the redox process involved could be tentatively rationalised as shown (Figure 4.18). Ascorbic acid can, in fact, be oxidised to its dehydro-derivative by a variety of oxidising agents such as the halogens iodine, bromine, and chlorine, quinones in acidic solution, iodate ion, phenolindophenol and other dyes, molecular oxygen in the presence of a suitable catalyst such as copper(II) ion at pH 5, and activated charcoal in aqueous ethanol. The reduction back to ascorbic acid may also be accomplished by several different reagents including hydrogen sulphide, hydriodic acid, cysteine and other thiols, and sodium dithionate. More recently, enzymes which catalyse this redox system have been investigated and characterised. The reaction of ascorbic acid with molecular oxygen is catalysed by the copper-containing enzyme ascorbic acid oxidase and the reduction of dehydroascorbic acid, most commonly by glutathione, by the enzyme DHA reductase.

The development of a satisfactory DHA synthesis has been hampered by the problems of over-oxidation and also by the limited stability of DHA in aqueous solution. To date, DHA has been obtained only as a viscous syrup or amorphous solid. A convenient preparative method involves oxidation by molecular oxygen using a catalyst of activated charcoal dispersed in aqueous ethanol. The DHA product is reasonably stable at low temperature (-10 °C). A dimeric form of DHA, bisdehydroascorbic acid (or BDHA), is also known. This substance is isolatable in crystalline form and may be obtained from DHA by dissolving the latter in nitromethane, boiling the solution and

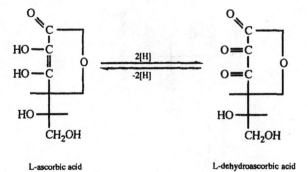

L-ascorbic acid · L-dehydroascorbic acid

Figure 4.18 *Early rationalisation of L-ascorbic acid/L-dehydroascorbic acid interconversion*

filtering off the precipitated, white dimer which may then be stored, under cool conditions, as a relatively stable solid.

Structure

As mentioned earlier the original structure determination work on L-ascorbic acid in the early 1930s led to the proposal that the first oxidation product was a neutral lactone whose aqueous solution underwent slow hydrolysis to a carboxylic acid, a process accompanied by a fall in pH and which could be followed by a gradual change in optical rotation. This lactone revealed no significant u.v. absorption above 225 nm, implying the absence of a carbon–carbon double bond as in ascorbic acid itself but the slow hydrolysis was accompanied by the development of a weak carbonyl signal at 290 nm. These observations were consistent with the equilibria shown (Figure 4.19). Later work, however, refined these original suggestions and, by the late 1970s, the structures of both DHA and BDHA were more precisely known. The monomer was shown to exist predominantly as a bicyclic, hemiketal species (the hemiketal involving the C-6 hydroxyl) with the dimer arising from loss of two water molecules between the C-2 and C-3 hydroxyls on adjacent monomers (Figure 4.20). The structure of DHA was confirmed essentially by spectroscopic techniques and, in particular, ^{1}H and ^{13}C n.m.r. The ^{1}H n.m.r. spectrum of DHA (in D_2O) clearly shows that H-6 and H-6' (4.2δ p.p.m.; integration 2), H-5 (4.6δ p.p.m.; int. 1) and H-4 (4.75δ p.p.m.; int. 1) constitute an ABMX system. Correct assignment of these signals is straightforward because double irradiation at 4.6δ (*i.e.* H-5) results in the multiplet at 4.2δ

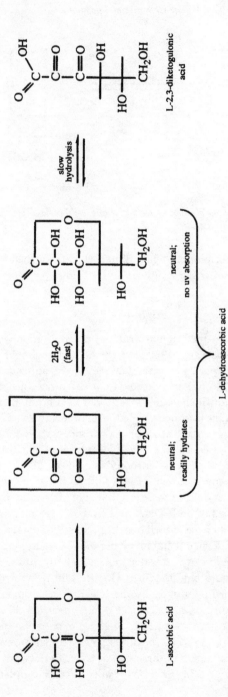

Figure 4.19 *Extended rationalisation of oxidation and lactone ring opening*

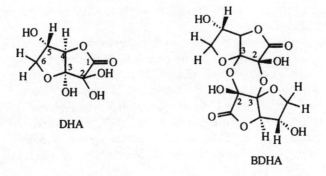

DHA

BDHA

Figure 4.20 *Dehydroascorbic acid monomer (DHA) and dimer (BDHA)*

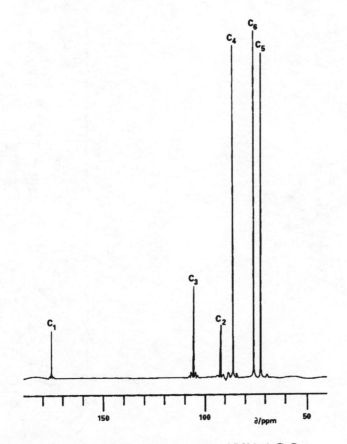

Figure 4.21 *Proton-decoupled* ^{13}C *n.m.r. spectrum of DHA (in* D_2O*)*

collapsing to a very obvious pair of 'roofed' doublets, the AB pattern for H-6 and H-6'.

The proton-decoupled ^{13}C n.m.r. spectrum shows six signals (Figure 4.21). Off-resonance decoupling reveals three singlets (173, 106, and 91δ p.p.m.), two doublets (87 and 73δ p.p.m.), and one triplet (76δ p.p.m.). Selective proton decoupling at the known H-4 and H-5 frequencies enabled the C-4 and C-5 doublets to be distinguished. The low-field singlet at 173δ is clearly the lactone carbonyl but it is interesting that the C-2 and C-3 carbon signals (91δ and 106δ respectively) experience considerably less deshielding which suggests

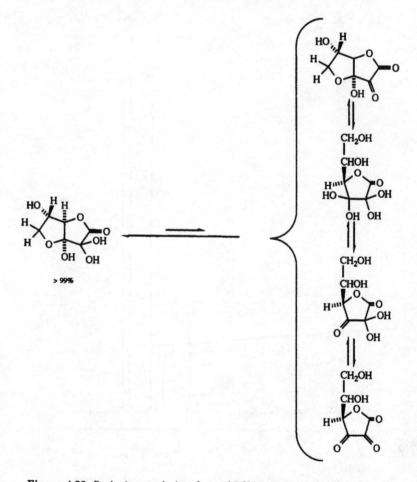

Figure 4.22 *Predominant and minor forms of DHA in aqueous solution*

that neither is present as a free carbonyl group. At the same time, C-6 is markedly more deshielded than in its L-ascorbic acid precursor (76δ compared with 63δ) indicating the involvement of the C-6 hydroxyl in hemiketal formation. Thus the *gem*-diol nature of C-2 and the hemiketal structure at C-3 are both supported by these data.

In aqueous solution, DHA exists predominantly (>99%) as the bicyclic hemiketal (Figure 4.20) but small quantities of several other forms are believed to be present although there is not a great deal of hard evidence for them (Figure 4.22).

Properties

DHA in its syrupy or amorphous solid form is stable for months at low temperature. In water, in which DHA readily dissolves, the rate of hydrolysis to 2,3-diketogulonic acid is very pH dependent, being relatively slow in cold solutions at pH 2–4 but more rapid above or below this range. Interestingly, despite the appearance of a weak u.v. carbonyl absorption on hydrolysis, ^{13}C n.m.r. studies of the 2,3-diketogulonic acid suggest that it exists largely with its C-2 and C-3 carbonyls in *gem*-diol form (Figure 4.23).

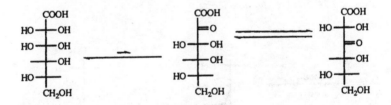

Figure 4.23 *Structures of hydrolysed DHA in aqueous solution*

The stable dimer BDHA dissociates in aqueous solution to DHA but is not, in fact, particularly water soluble.

DHA reacts with *o*-phenylenediamine to give products which depend on the stoichiometry (Figure 4.24).

With phenylhydrazine, DHA (and ascorbic acid itself, which is initially oxidised to DHA) yields a red, crystalline 2,3-bisphenylhydrazone (or osazone), m.p. 197 °C, whose structure is stabilised by intramolecular hydrogen-bonding (Figure 4.25). These bishydrazone derivatives are of interest in their own right in that they can be converted, by rearrangement, into pyrazole ring structures and, by

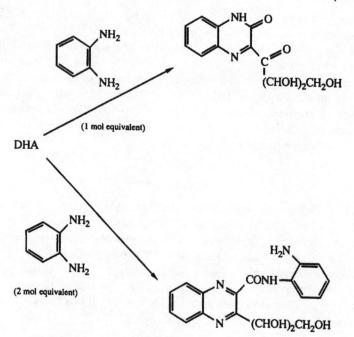

Figure 4.24 *DHA–o-phenylenediamine reaction*

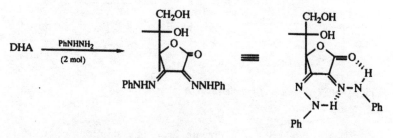

Figure 4.25 *Bishydrazone of L-ascorbic acid*

reduction, into 2,3-diamino-2,3-dideoxy derivatives of L-ascorbic acid which, in turn, react with a variety of aldehydes to produce imidazolines (Figure 4.26). Indeed, rearrangement to the yellow pyrazoledione (m.p. 210 °C) can sometimes occur simply on recrystallisation of the bishydrazone, thus explaining the early uncertainty over the nature of the phenylhydrazine condensation product.

Another interesting reaction of DHA involves the formation of a red

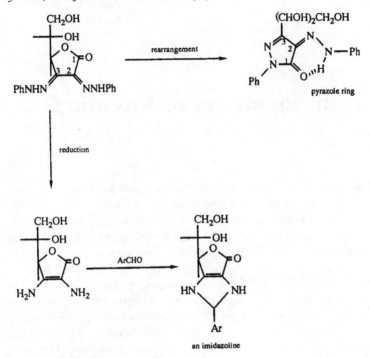

Figure 4.26 *2, 3-Bishydrazone modification*

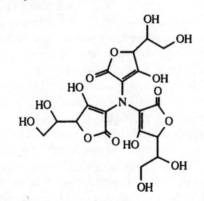

Figure 4.27 *A tris(2-deoxy-2-L-ascorbyl)amine*

compound with α-amino acids, a reaction that has been used for their identification following chromatographic separation. The structure of these coloured derivatives has now been elucidated as a tris(2-deoxy-2-L-ascorbyl)amine (Figure 4.27).

Biochemistry of Vitamin C

It is a reasonable assumption that no one reading this book will have been ignorant of the antiscorbutic nature of vitamin C. Those who have studied even a little biology may well further appreciate that the vitamin's wound healing and growth promoting properties are due to its participation in the synthesis of fibrous connective tissue, specifically in promoting the post-translational hydroxylation of proline and lysine residues in collagen, the most abundant protein of the Animal Kingdom. In this chapter on the biochemistry of vitamin C this process, still far from being fully understood, whereby, paradoxically, the reducing substance ascorbic acid is necessary to bring about the oxidation of proline and lysine, will constitute a major topic. However, since the early years of vitamin C biochemistry, starting with the controversy surrounding its discovery and structure elucidation, in addition to its role in amino acid metabolism, an ever increasing sphere of influence and involvement has come to light including aspects of immunology, oncology, digestion and absorption, endocrinology, neurology, detoxification, and cataract prevention.

Although it appears that only a small number of species of higher organisms is incapable of its synthesis, because this band of dependants includes *Homo sapiens*, it is perhaps not surprising that most of what is known of its biochemistry relates to mammals. The vast majority of scientific papers devoted to the metabolic role of ascorbic acid report findings from work with the guinea-pig and the laboratory rat, the title 'vitamin' only strictly applying to the former. Like us, the guinea-pig is able to synthesise little or no ascorbic acid whereas the laboratory rat appears to be self-sufficient. From our weaning (human milk contains up to 5 mg% ascorbic acid depending on the mother's diet) to the grave, we depend for our vitamin C mainly on fruit and vegetables, unless a synthetic supplement is taken. Despite the notorious abundance of the vitamin in some plant materials and its ubiquity in chlorophyll-containing plants generally, very little is to be learned from botanical or biochemical texts as to its purpose in plants.

Beginning with an account of ascorbic acid biosynthesis, this chapter will conclude with a consideration of its catabolism and excretion, inevitably from a mammalian viewpoint. Sandwiched in between the vitamin's origins and destruction will be a catalogue of, as yet, frustratingly only partly understood biochemical involvements in the host of biological phenomena mentioned above. Some order may be apparent in all this in that it is possible to categorise some of its known biochemical reactions as oxidations (more specifically hydroxyla-tions), reductions (for example, the protection of the sulphydryl group of the easily oxidised tripeptide glutathione and the elimination of potentially harmful oxidising free radicals), and redox reactions (concerned with electron transfer and membrane potentiation through the establishment of a proton gradient). Ascorbic acid itself is a reducing agent and therefore cannot directly promote oxidations. However, inside the living cell the vitamin can occur in several different forms, various pairings of which constitute redox couples, capable of bringing about both oxidations and reductions of compo-nents of other redox couples, depending on their relative redox potentials.

The various physiologically active forms of vitamin C are as shown in Figure 5.1 but it is by no means certain that all the interconversions shown take place *in vivo*. Indeed our continued inability to pinpoint a definitive biochemical role for this most vital of substances is a source of continuing frustration.

BIOSYNTHESIS

Most living organisms probably can convert D-glucose into L-ascorbic acid and it is important to realise that the conversion comes about *not* through epimerisation but through turning a D-structure upside down and then calling it an L-structure, – the carbon chain inversion described in Chapter 3. One has to remember that the open-chain Haworth representation of D-glucose shown in Figure 5.2 is so called (D) because the hydroxy-group on the penultimate carbon atom (C-5 in this case) projects to the right.

Provided its vulnerable aldehydic C-1 can be protected, D-glucose can be modified by oxidation of C-6 to D-glucuronic acid as shown in Figure 5.3.

If the C-1 aldehyde group is now reduced to a primary alcohol

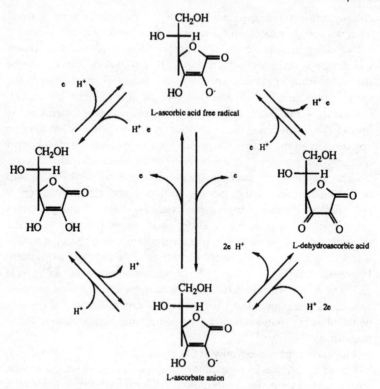

Figure 5.1 *Interconversions of various forms of vitamin C*

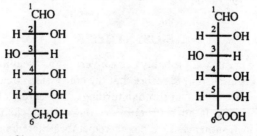

Figure 5.2 *D-Glucose*   **Figure 5.3** *D-Glucuronic acid*

group, then one has the structure shown in Figure 5.4. Convention now dictates that the highest-priority functional group (in this case the carboxylic C-6) should be placed at the top and renumbered C-1. Hence if the page is turned upside down, or the drawing is inverted, then an L-compound is apparent, as seen in Figure 5.5. This compound is a derivative of the aldohexose L-gulose and is therefore called L-gulonic acid. Its cyclisation and oxidation gives rise to the formation of L-ascorbic acid as shown in Figure 5.6 and this can be redrawn in the more usual form given in Figure 5.7.

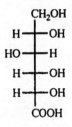

Figure 5.4 *Product of reaction at C-1 of glucuronic acid*

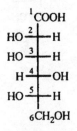

Figure 5.5 *L-Gulonic acid (Figure 5.4 inverted)*

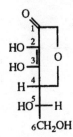

Figure 5.6 *L-Ascorbic acid*

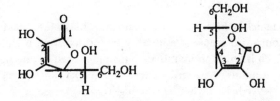

Figure 5.7 *Different ways of representing L-ascorbic acid*

It is thought that all chlorophyll-containing plants and, in the case of the spermatophytes, their germinating seeds can synthesise ascorbic acid and one proposed route is shown in Figure 5.8.

Some authorities suggest that a more important pathway in plants involves D-galacturonic acid which, by reduction and lactonisation, would produce, as an alternative to L-gulonolactone, L-galactourono-

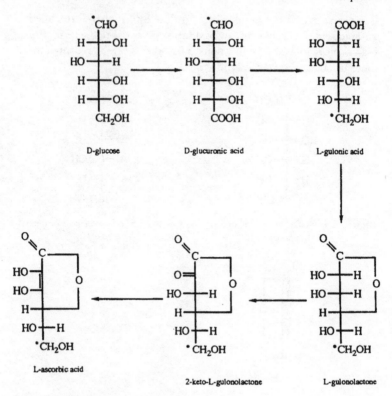

Figure 5.8 *Biosynthesis of L-ascorbic acid*

lactone which could be converted into L-ascorbic acid. As D-galacturonic acid would presumably be formed from D-galactose, a sugar which is found in mammals but not, at least in any significant amount, in plants, it is hard to accept that this can be a significant process. Nevertheless, whether starting from D-glucose or D-galactose, it is clear that inversion is necessary to produce L-ascorbic acid. At variance with this is the experimental observation that some plants, including ripening strawberries, convert D-glucose to L-ascorbic acid with no inversion. Using D-glucose labelled with tritium (^{3}H) at C-6 L-ascorbic acid was produced with most of the label still associated with C-6. Some recent genetic engineering, in which the bacterium *Erwinia herbicola* was able to express a reductase enzyme derived from a gene belonging to a species of *Corynebacterium*, introduces two more intermediates, D-gluconic acid and 2,5-diketogluconic acid, in the pathway shown in Figure 5.9. In this case no inversion is necessary to explain how D-glucose could be transformed into L-ascorbic acid and for the present it would seem appropriate to keep an open mind and assume that some plants invert via D-glucuronic acid or D-galactouronic acid to produce L-gulonic acid and others proceed via D-gluconic acid which needs no inversion if 2,5-diketogluconic acid is a further intermediate.

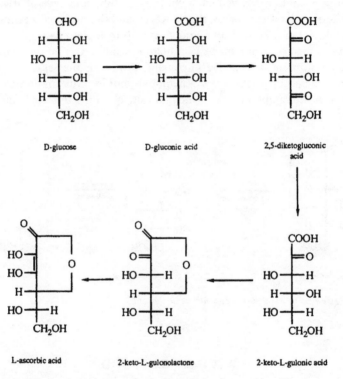

Figure 5.9 *An alternative route for the biosynthesis of L-ascorbic acid*

Most of the research into animal ascorbic acid synthesis has involved the laboratory rat and the findings support the idea of full C-1/C-6 inversion between D-glucuronate and L-gulonate. Apart from primates (including *Homo sapiens*), a few other mammals, and some birds, fish, and insects, it is thought that the vast majority of the members of the Animal Kingdom convert D-glucose into L-ascorbic acid via D-glucuronic acid, L-gulonic acid, L-gulonolactone, and 2-keto-L-gulonolactone as intermediates. Why this ragbag minority of animal species lack the ability, apparently due to their ancestors losing the genetic material necessary to synthesise the last enzyme in the chain, L-gulonolactone oxidase, is a mystery. This genetic lesion blocks the final reaction in the sequence given in Figure 5.10. It has been suggested that this careless slip occurred 25 million years ago and has led to man sharing this apparent embarrassment with the other primates, the humble guinea-pig, the Indian fruit bat, some birds, including the exotic sounding red-vented bulbul, some fish, and several insects including the desert locust, the silkworm, and the *Anthonomus* genus of beetles. It may be that all phytophagous (green plant eating) insects have a dietary requirement. Withholding vitamin C from the locust's diet results in an abortive moult and death. In the self-sufficient, synthesis occurs in the livers of mammals, but in birds, and probably reptiles and amphibians too, the main site of synthesis is the kidney.

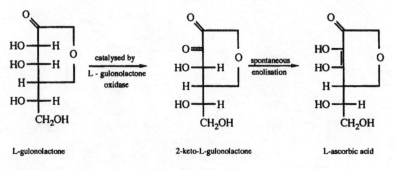

L-gulonolactone   2-keto-L-gulonolactone   L-ascorbic acid

Figure 5.10 *Final step in biosynthesis of L-ascorbic acid*

VITAMIN C IN FOOD

As far as *Homo sapiens* is concerned, milk is the only animal product that provides a significant amount of the vitamin (1–5 mg per 100 g) and, although there is some in liver, the best sources are fresh fruits (particularly citrus fruits, tomatoes, and green peppers), baked potatoes (17 mg per 100 g), and leafy vegetables. Some fruit, like guavas (300 mg per 100 g) and blackcurrants (200 mg per 100 g), are particularly rich in the vitamin but they contribute little to normal Western dietary intake. Table 5.1 gives a fairly comprehensive list of the vitamin C content of a wide variety of common foods.

Although fruit and vegetables initially contain large amounts of

Table 5.1 *Vitamin C content of some common fruits and vegetables*

Fruit or vegetable	Ascorbic acid content/mg per 100 g
Rosehip	1000
Blackcurrant	200
Kale	186
Green pepper	128
Horse radish	120
Broccoli	113
Brussels sprouts	109
Watercress	79
Cauliflower	78
Strawberry	59
Spinach	51
Oranges/lemons	50
Cabbage	47
New potato	30
Pea	25
Old potato	8
Carrot	6
Apple	6
Plum	3

These figures vary of course from fruit to fruit depending on a wide variety of parameters such as the source, soil conditions, prevailing weather and many others.

ascorbic acid, the culinary treatment which they receive often serves to reduce the amount present before it reaches the consumer. Thus, plant material rich in ascorbic acid also contains ascorbic acid oxidase which may normally be inactive or contained within vesicles, since fine chopping of such material is known to increase the enzyme's activity.

Another enzyme which will cause loss of ascorbic acid in plant material is phenolase. This is the substance which helps to produce the browning of fruit such as apples when polyphenolic species are oxidised by oxygen from the air. The enzyme functions with oxygen and ascorbic acid and reduces *ortho*-quinones back to *ortho*-diphenols. This results in the formation of dehydroascorbic acid, which is rapidly converted to 2,3-diketogulonic acid. The process is catalysed by copper(II) and other transition metal ions, which therefore accelerate the loss of ascorbic acid from vegetables and fruit, when for example they are cooked in copper or iron containers. Of course the major factor in the removal of vitamin C from cooked vegetables is simple dissolution in the cooking water.

There have been several studies of the rate of destruction of vitamin C resulting from the processes of boiling and cooking. Most of these have been on model systems and generally the assumption made is that the loss of vitamin C follows a first-order process and that the Arrhenius equation (2) adequately describes the effect of temperature on the rate of destruction, where k = first-order rate constant for the loss of the vitamin, A = pre-exponential factor, E_a = activation energy for the process, R = gas constant, and T = temperature/K.

$$k = Ae^{-E_a/RT} \qquad (2)$$

Experiments have shown that by plotting $\ln k$ against $1/T$, the resulting straight line allows the calculation of the activation energy. Typical values lie in the range 80–120 kJ mol^{-1}. However, plants constitute a very complex system and many factors contribute to the destruction of the vitamin such as micro-organisms and/or natural enzymes. Thus, activation energies provide only a very rough guide to the effect of temperature on vitamin destruction. Furthermore, an additional parameter is the presence or absence of air. The situation is further complicated if the reaction is not first-order. For example, there is evidence to suggest that the decomposition of ascorbic acid in stored juices is zero-order. It is difficult to see why there is this difference in kinetic behaviour, but no doubt the mechanism of the vitamin destruction process depends crucially on the nature of the system. It is perhaps worth noting that vegetables cooked in a microwave oven retain the vitamin C content more than those cooked under conventional conditions. This seems largely due to the much smaller amount of water which is used in microwave cooking, though the shorter time required must also be a factor. Hence culinary losses of the vitamin from vegetables can be minimised not only by avoiding over-long boiling in copper pans, but by cooking fruit and leaves whole.

OXIDATION AND HYDROXYLATION

Ascorbic acid is now well established as an essential factor in many hydroxylation reactions of the type RH + O→ROH. On the face of it this seems a paradoxical role for a reducing substance but not if one treats the vitamin as a redox couple, *e.g.* ascorbic acid/dehydroascorbic acid (H_2A/A) which will undergo cycling, like the cytochromes. After all, what is really meant by reference to, say, cytochrome c in the context of its role in the mitochondrion is cytochrome c (Fe^{II})/ cytochrome c (Fe^{III}) because, in helping to transfer electrons from metabolites towards oxygen, the cytochrome molecule continually alternates or cycles between being reduced (Fe^{II}) and oxidised (Fe^{III}). In hydroxylation, the reactive oxygen is generated in a so called mono-oxygenase catalysed reaction which can be expressed as:

$$XH_2 + O_2 \rightarrow X + H_2O + O \qquad (3)$$
$$O + RH \rightarrow ROH \qquad (4)$$

While XH_2 could be identical with ascorbic acid itself, it is more likely to be another reduced substance, *e.g.* $NADH_2$, $NADPH_2$, α-ketoglutarate, *etc.*, with ascorbic acid and a cytochrome playing a cycling redox role as shown in Figure 5.11. In some cases ascorbic acid may act as substance XH_2 and the reaction could be simplified to:

$$H_2A + O_2 + RH \rightarrow A + H_2O + ROH \qquad (5)$$

Ascorbic acid is known to be involved in the metabolism of several amino acids, leading to the formation of hydroxyproline, hydroxylysine, noradrenalin (norepinephrine), serotonin, homogentisic acid, and carnitine.

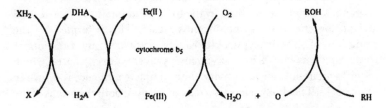

Figure 5.11 *Ascorbate redox recycling contribution to oxidase/hydrolase activity*

Hydroxyproline and hydroxylysine are constituents of collagen, the main protein of fibrous connective tissue in animals, and occur in hardly any other proteins. Contributing nearly one-third of total mammalian body protein, collagen occurs in tendons, ligaments, skin, bone, teeth, cartilage, heart valves, intervertebral disks, cornea, and lens in addition to its general tissue distribution as extra-cellular framework. There are no codons for hydroxyproline (Hyp) and hydroxylysine (Hyl) and, when collagen is synthesised, proline (Pro) and lysine (Lys) are hydroxylated post-translationally on the growing polypeptide chain. About half the proline residues are hydroxylated (mainly on C-4 but with some at C-3) and, while total proline (Pro and Hyp) accounts for about 27% of the amino acid residues, the single most abundant is glycine (Gly) at 26% with total lysine (Lys and Hyl) contributing only about 4% of the residues. The proline and lysine hydroxylase requires ascorbic acid, iron(II), and α-ketoglutarate (αKG) as the co-oxidant. It has been speculated that the reaction may proceed via the intermediate formation of a peroxyglutarate reacting with proline or lysine as shown in Figure 5.12 for proline. For hydroxyproline formation, the enzyme is specific for proline residues, occupying the third position in a triplet of the type Gly X Pro *followed* by a glycine (Gly). It seems that the role of ascorbic acid is in maintaining the hydroxylase's iron co-factor in a reduced (Fe^{II}) state at the active site. The subsequent triple helix quaternary state of physiologically effective collagen can only be accomplished if the requisite proline and lysine residues have been hydroxylated. Collagen formed in the absence of vitamin C is unable to form proper fibres resulting in the skin lesions and blood vessel fragility so characteristic of scurvy. Another protein which depends on vitamin C, and which also contains hydroxyproline, is the plasma complement protein (C-1qu) which constitutes an important part of the guinea-pig's defence against pathogens.

The amino acids phenylalanine and its hydroxylated derivative, tyrosine, are both catabolised in the livers of animals to fumaric acid and acetoacetic acid via homogentisic acid. This is formed by the oxidation of 4-hydroxyphenylpyruvate, catalysed by the copper containing enzyme 4-hydroxyphenylpyruvate dioxygenase, which requires vitamin C for its activity. The complete sequence is shown in Figure 5.13. The dioxygenase is so called because both the atoms of the

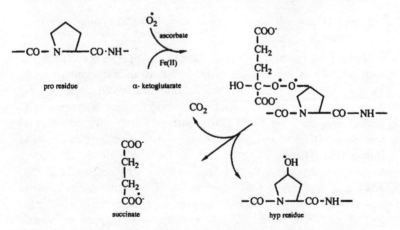

Figure 5.12 *Possible mechanism for post-translational hydroxylation of proline*

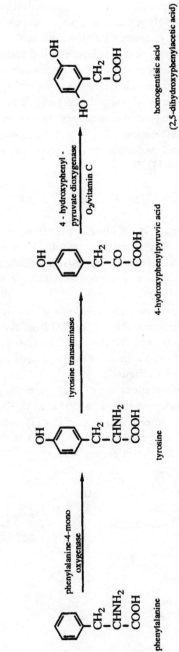

Figure 5.13 *Metabolism of phenylalanine and tyrosine*

oxygen molecule are incorporated into the product. It should be noted that, although decarboxylation also takes place, no reference to that occurs in the name of the enzyme. Homogentisic acid is a landmark metabolite in biochemistry in that the discovery and understanding of the inherited disorder alcaptonuria, which arises when an individual lacks homogentisate oxidase, resulted in the coining of the term 'inborn error of metabolism', scores of which have been discovered subsequently. At the beginning of the sequence the hydroxylation of phenylalanine (Phe) to tyrosine (Tyr), accompanied by the co-oxidation of the folic acid derivative tetrahydrobiopterin (THB) to produce dihydrobiopterin (DHB),

$$Phe + O_2 + THB \rightarrow DHB + Tyr + H_2O \qquad (6)$$

and the subsequent reconversion of DHB to the tetrahydro-form, may be vitamin C-dependent. The congenital inability to convert phenylalanine into tyrosine afflicts about 1 in 20 000 individuals and is known as phenylketonuria, an inborn error of devastating consequence if untreated.

Tyrosine is the precursor for the synthesis of the hormone and neurotransmitter noradrenalin which is formed from dopamine (3,4-dihydroxyphenylethanolamine) by the vitamin C-requiring dopamine β-mono-oxygenase (also called dopamine-β-hydroxylase) in the adrenal medulla as shown in Figure 5.14.

Both the above hydroxylations seem to involve the recycling of tetrahydrobiopterin which may require ascorbic acid. Recently it has been suggested that dopamine-β-hydroxylase (DBH) works in tandem with semidehydroascorbate reductase (SDR) in order to recycle the vitamin and oxidise $NADH_2$ as in Figure 5.15.

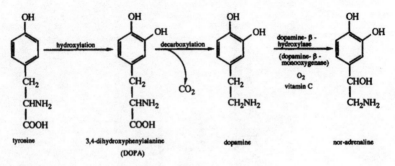

Figure 5.14 *The formation of noradrenalin*

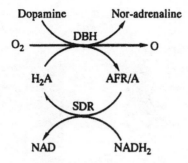

Figure 5.15 *Ascorbate recycling in noradrenalin synthesis*

What happens to the released, highly reactive and therefore highly dangerous, oxygen atom is not clear. The hydroxylation and decarboxylation of tyrosine to form dopamine is paralleled in the pineal gland for the other aromatic amino acid, tryptophan, to form the vasoconstrictor and neurotransmitter 5-hydroxytryptamine, better known as serotonin. The initial hydroxylation step catalysed by tryptophan-5-mono-oxygenase (or tryptophan hydroxylase), is thought to require vitamin C and is shown in Figure 5.16.

The co-substrate for the mono-oxygenase is tetrahydrobiopterin (THB) and again it is likely that ascorbic acid functions to restore this substrate from the oxidised dihydrobiopterin (DHB) as in Figure 5.17.

Lysine is converted into carnitine through a chain reaction driven at two points by dioxygenases (or hydroxylases) that act on γ-butyrobetaine and trimethyl-lysine and that require vitamin C for full activity. Carnitine is essential for the transport of energy-rich fatty acids from the cytoplasm to the mitochondrial matrix where they are catabolised by β-oxidation to acetate.

Other oxygenase enzyme systems responsible for neurotransmitter and hormone synthesis, and dependent on the presence of vitamin C, are the copper-containing peptidyl glycine α-amidating mono-oxygenases (PAMs), recently discovered in the pituitary and adrenal glands. *C*-Terminal glycine peptides, which are widely distributed in endocrine and nervous tissue, and may be represented as AA_n-Gly-COOH, are converted by PAMs into AA_n-$CONH_2$ in a reaction similar to that brought about by dopamine β-mono-oxygenase. In this form they are thought to be more resistant to enzymic attack and have an enhanced ability to bind with their receptors. Examples include melanocyte stimulating hormone and thyrotropin releasing hormone.

One important function of mammalian liver is to bring about

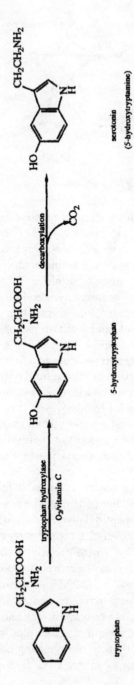

Figure 5.16 *The formation of serotonin*

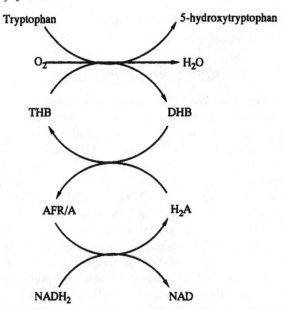

Figure 5.17 *Ascorbic and hydrobiopterin recycling in serotonin synthesis*

biotransformation of xenobiotics, *i.e.* drugs, poisons, and abnormal metabolites. A significant role is played by oxidising enzymes associated with the microsomal fraction, notably a cytochrome called P450 which is a mixed function oxidase (MFO). This very broad-specificity enzyme, which helps dispose of ethanol, benzene, tetra-chloromethane, and polychlorinated biphenyls (PCBs) to name but four notorious toxins, may depend for its action on the presence of vitamin C since, in a deficiency state in the guinea-pig, the level of this enzyme is markedly decreased. Furthermore, the administration of some of these compounds increases the rate of metabolism and the dietary requirement of the vitamin in experimental animals.

Another hydroxylation role for vitamin C in the hepatic microsomal fraction is the stepwise conversion of cholesterol to the bile acid, cholic acid, via 7α-hydroxycholesterol, 3α,7α-dihydroxycoprostane and 3α,7α,12α-trihydroxycoprostane. Also, in lipid metabolism, conventional fatty acids with an even number of carbon atoms are α-oxidised by a mono-oxygenase and subsequently decarboxylated to form an odd-numbered carbon derivative and both these steps appear to require ascorbic acid. As the initial α-oxidation is brought about by a mono-oxygenase, it is possible that all such MFOs may require the vitamin.

REDUCING PROPERTIES

Although life as we know it is usually dependent on oxygen, it is possible to have too much of a good thing. In excess, in the wrong form, or in the wrong place, oxygen is a potent poison. Particularly harmful are the very aggressive species of oxygen and the oxidising radicals, *e.g.* the superoxide anion and the hydroxyl radical, collectively known as the reactive oxidants which can inflict serious damage on the lipid components of cellular membranes by peroxidation. The protective anti-oxidant role of vitamin E and the essential fatty acids is well recognised. However, they are fat-soluble and it appears that their intra-membrane function is taken over by ascorbic acid on the surfaces of membranes. Here, and in other aqueous parts of the cell, vitamin C assists another water soluble anti-oxidant, the tripeptide glutathione, in mopping up potentially dangerous oxidising radicals.

Glutathione is a tripeptide with the primary structure γ-Glu-CySH-Gly and exists either as such (GSH) or as the oxidised dimer (GSSG). In the GSH form, it plays a vital role in maintaining various key proteins in the reduced state necessary for such varied functions as the structural integrity of red blood cells and the transparency of the lens of the eye. There is a natural tendency for GSH to oxidise to the non-functional GSSG form and it has been suggested that ascorbic acid either prevents or reverses this undesirable reaction. This may not be a direct effect as GSH is also known to be maintained by a supply of $NADPH_2$, generated by the pentose phosphate pathway in which glucose-6-phosphate is degraded to ribulose-5-phosphate. This pathway is known in many plants and micro-organisms as well as in the lens, red blood cells, adrenal gland, adipose tissue (including lactating mammary gland), and liver of animals and there is recent evidence of its stimulation by ascorbic acid. Paradoxically, there is some suggestion that one of the functions of glutathione is to maintain ascorbic acid in the reduced state!

Describing vitamins E and C as having identical anti-oxidant roles in lipid and aqueous cellular locations respectively is almost certainly an over-simplification since they have been shown to interact synergistically and it may be that at the lipid/aqueous interface, ascorbic acid protects vitamin E, or is involved in its restorative reduction after a successful attack on an oxidising free radical.

The reducing property of ascorbic acid also 'assists' another vitamin, folic acid (Figure 5.18). This is an essential co-factor in various one-carbon transfers: for example the methyl group originating from the essential amino acid methionine is required in the formation of a wide variety of compounds including purines, the pyrimidine thymine, the amino acid serine, choline, carnitine, creatine, adrenalin, and many others. In its functional state, folic acid must be in its most reduced tetrahydrofolate form and this is brought about and/or maintained by ascorbic acid.

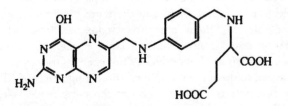

Figure 5.18 *Folic acid*

A particular problem in red blood cells is the tendency of the aggressive superoxide free radical to produce non-functional methae-moglobin (metHb) by oxidising the haem iron from Fe^{II} to Fe^{III}. This process is reversed by metHb reductase which involves cytochrome b_5 and ascorbic acid. The superoxide free radical is normally destroyed by superoxide dismutase (SOD), which some say depends on the vitamin for full activity. By destroying the superoxide radical, SOD prevents the subsequent formation of the very aggressive hydroxyl radical. However, and controversially, it is now claimed that ascorbic acid can by-pass this defence by reducing Fe^{III} to Fe^{II} which releases the hydroxyl radical from peroxide in the Fenton reaction. In plants, however, SOD and ascorbate, together with glutathione and $NADPH_2$, have been proposed as an integrated system of defence against the superoxide radical.

The well known phenomenon of ascorbic acid enhancing the absorption of iron from the intestine is probably due to it maintaining the element in the reduced, Fe^{II}, state in which it is more easily taken up across the mucosal membrane.

ELECTRON TRANSPORT

The redox properties of ascorbic acid have been used for many years in *in vitro* studies of electron transport by mitochondrial membranes.

Electron transport and oxidative phosphorylation are usually explained by two not very well linked models, one relating redox potential difference to free-energy change and the other linking exergonic electron transport to the establishment of a proton gradient. The mitochondrial electron transport chain can be represented as in Figure 5.19. Substrates like isocitrate, α-ketoglutarate, and malate of the citric acid cycle have a P:O ratio of 3, *i.e.* three ATP molecules are produced for each molecule of substrate oxidised by the removal of two atoms of hydrogen using NAD as a co-enzyme. This is because the electrons removed (with their protons) can pass all the way along the entire electron transport chain and there are three phosphorylation sites available, *i.e.* there are three redox reactions where the participating redox couples have a large enough redox potential difference (about 0.21 V) for sufficient free energy (about 40 kJ mol^{-1}) to be released to drive the phosphorylation of one mole of ADP.

If the substrate is succinate, also of the citric acid cycle, or an acylCoA formed during the β-oxidation of fatty acids, then, because their dehydrogenases are flavoproteins, they miss out on the first phosphorylation site and have a P:O ratio of 2.

Ascorbic acid will reduce cytochrome *c*, via the dye tetramethyl-*p*-phenylenediamine (TMPD); hence it misses out on the first two phosphorylation sites and therefore has a P:O ratio of unity.

In the 1960s, Mitchell showed how the energy released in electron transport is used to pump protons from the matrix side of the inner mitochondrial membrane to the cytoplasm side. The subsequent dissipation of the proton gradient, via 'gates' in the stalks of the knob-like projections of the inner membrane, activate ATP synthetase. Recently ascorbic acid has been implicated in such membrane potentiation by the establishment of proton gradients in the plasma membrane vesicles extracted from the soybean (*Glycine max*).

Modern representations of electron transport in organelles like mitochondria and chloroplasts envisage some carriers being static or fixed in a specific location within the membrane, others being more loosely associated with membrane surfaces, and yet others being mobile within the confines of the membrane. It is tempting to suggest that vitamin C, rather than being the artificial electron donor used to reduce cytochrome *c in vitro*, may be one more link in the electron transport chain in intact living mitochondria and chloroplasts. That vitamin C can traverse membranes is obvious from its occurrence in varying, and in some cases high, concentrations in the cells of various tissues, *e.g.* the adrenal gland and brain. Although inside the cells of

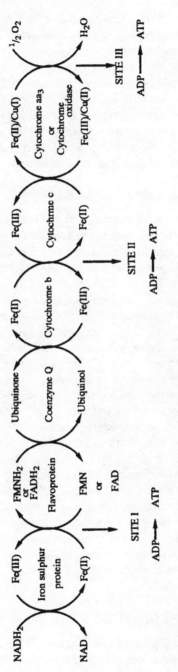

Figure 5.19 *The mitochondrial electron transport chain and oxidative phosphorylation sites*

such tissues ascorbic acid (H_2A) may predominate, it is known that the membranes are permeable only to dehydroascorbic acid (A). If the cells of the adrenal gland accumulate H_2A by oxidising it outside, thus allowing A to enter, and then reconstitute H_2A by intracellular reduction, then it seems reasonable that some ascorbate redox couples (H_2A/A or H_2A/AFR) could be involved in electron transfer between lipid- and water-soluble components in cells and/or organelles. Some recent work with plants suggests that ascorbic acid is implicated in establishing a proton gradient subsequently used to transport material across cell membranes and thus may be involved in growth regulation.

TISSUE LOCATION

The specific involvement of vitamin C in hydroxylation to produce collagen, serotonin, and noradrenalin are all animal processes and yet it must be very important in its main site of synthesis, *i.e.* chlorophyll-containing plants. Despite its very high concentration in some plant tissues and its rapid rate of synthesis in germinating seeds, very little seems to be known of its involvement in metabolic events, apart from its apparent necessity for xanthophyll synthesis, the desaturation (oxidation) of certain fatty acids, and its possible involvement in translocation mentioned above. A clue to its metabolic role in animals could be determined from a study of tissue analysis. The mammalian tissues which have been analysed, in decreasing order, are the adrenal gland (55 mg%), pituitary and leucocytes (white blood cells), brain, lens and pancreas, kidney, spleen and liver, heart muscle, milk (human 3 mg%, bovine 1 mg%), and, finally, plasma (1 mg%). In most of these tissues, common functions of vitamin C will be in maintaining structural integrity through collagen synthesis. The higher levels in some of the more vital organs may be to protect them against dietary deficiencies presumably because, although distressing, a scorbutic skeletal joint is far less damaging than a scorbutic brain. More likely, however, is that these elevated levels reflect more specialised roles, *e.g.* the synthesis of hormones and neurotransmitters in the adrenal and brain, contributing to the immune response in spleen and leucocytes, promoting the pentose phosphate pathway in liver, and maintaining transparency in lens and cornea.

INTAKE, EXCRETION, AND CATABOLISM

From a human dietary point of view, vitamin C is seen primarily as being required to prevent scurvy with as little as 10 mg per day being effective in this respect but with a recommended daily allowance being 30 mg per day in the UK. By contrast the laboratory rat synthesises an amount that has been calculated to be the equivalent of 2000 mg (2 g) per day in the human adult! Clearly the rat and other animals, including man, and plants that contain no collagen, use ascorbic acid for a purpose other than preventing scurvy. The school, currently out of favour (see Chapter 6) that recommends megadoses (1–10 g per day) may have a point. One of the arguments raised against the megadose recommendations is that the adult human body accumulates a limited amount (normally 2–3 g) of the vitamin, possibly rising to a maximum of 4 g. At this point plasma levels reach about 1.4 mg% and any intake in excess of the daily rate of metabolism (5–20 mg per day) is absorbed but promptly excreted unchanged through the kidneys. A recent claim that no more than 500–600 mg daily can be absorbed further weakens the case against the megadose intake. That which is metabolised is broken down by the liver, and to some extent also by the kidney, in a series of reactions, culminating in the formation of oxalic acid, which is excreted in the urine. The catabolism of ascorbic acid in animals may follow the pathway shown in Figure 5.20. It has been suggested that as ascorbic acid oxidase is active in dormant seeds its inhibition, or rather the accompanying rise in levels of ascorbic acid, may be a trigger in breaking dormancy.

While something is known about the biosynthesis and degradation of vitamin C, being able to ascribe a specific co-enzyme type role remains tantalisingly elusive. Its well known reducing property makes it an excellent co-substrate in mono-oxygenase reactions producing hydroxylated amino acids and catecholamines; its reducing properties also protect cells by scavenging for oxidising free radicals and in protecting other anti-oxidants, like vitamin E. Its chelating and/or reducing properties help in the intestinal absorption of iron. It has been speculated that it may function as a cycling redox couple in electron transport and membrane potentiation phenomena and may have the same sort of status presently accorded to cytochrome c. At present the only general role at the metabolic level that can be proposed is that it may be the optimal, but not in all cases the only, factor needed to keep many iron- and copper-containing enzymes in their reduced state, where they appear to be most effective. Three world conferences have been devoted to vitamin C, the latest in 1986 which included sessions on neurochemistry, health findings from epidemiology, health and disease, biochemistry and immunology, xenobiotics, analysis, metabolism, and safe levels of ingestion. This conference was undoubtedly,

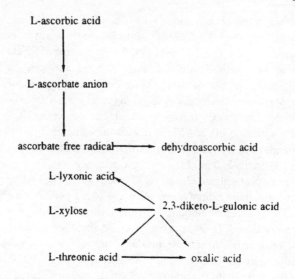

Figure 5.20 *The catabolism of ascorbic acid*

and not surprisingly, attended mainly by medical scientists and
scarcely a mention was made of the world's main sphere of biosynthesis
and use – the Plant Kingdom. Our understandable interest, encouraged by industrial funding, into animal, *i.e.* mammalian, but
particularly human, aspects may be the worst place to seek to unravel
the continuing mystery of whether or not there is one primary
biochemical function for this simple molecule. Our parochial passion is
generated by our lacking L-gulonolactone oxidase. A gene deleted
from our ancestors 25 million years ago has made man and the other
primates, together with some birds, bats, and beetles and, of course, the
humble guinea-pig, obligate vegetarians.

Chapter 6

Medical Aspects of Vitamin C

It could be argued that the discovery and application of vitamin C, albeit unidentified and masquerading as citrus fruits, in preventing scurvy in Spaniards venturing across the Atlantic and Pacific Oceans in the sixteenth and seventeenth centuries marked the beginning of dietetics as a branch of modern medicine. Lemon juice was also included in the provisions of the vessels of the Dutch East India Company from the seventeenth century and, rather more surprisingly, citrus juices seem to have been issued in some English ships during the early part of the eighteenth century, but then disappeared from the record until James Lind's remarkable clinical trial, discussed in Chapter 2, which culminated in his recommendation of 1753 that all sailors be issued with a daily ration of lemon juice. The almost certainly deliberate withholding of antiscorbutic foodstuffs in order to keep the seamen weak and controllable by the East India Company, among others, represents a less than savoury aspect of human society. Equally squalid were the personal jealousies which caused the British Navy to delay implementing Lind's recommendations for 42 years, a delay that was nothing less than a scandal in medical history and which was halted only by a mutiny, one of whose demands was 'a sufficient quantity of vegetables'. Over the succeeding two hundred years, the active ingredient of health-maintaining fruits and vegetables has been identified, prescribed in daily doses ranging from a few milligrams to many grams, implicated in preventing and curing (or in some cases causing) almost every major disease state known to man and laboratory mammals, and formed the basis of a huge international pharmaceutical business through its chemical synthesis and formulation into tablets. Yet, as we have seen, this simple, bitter-sweet, sugar-like substance still hides the secret as to its exact physiological role despite its successful application in various, apparently unrelated, medical situations. For hundreds of years it has been used to cure scurvy and in the past two years it has brought about, in some patients, the remission of autoimmune thrombocytopaenia, with a combination of a degree of faith and a lack of pharmacological understanding almost unequalled in the medical profession. Such therapeutic uses generally involve the oral prescription of hundreds or thousands of milligrams daily which is far in excess of the relatively tiny amount needed to prevent the onset of scurvy. Dietary intakes by the non-sick who follow, or in many cases exceed, the recommended daily dietary allowance (RDA) are also large compared with the antiscorbutic intake and there is concern expressed in some quarters that these high intakes may cause undesirable side effects. While in some disease states it would appear that vitamin C intake should be kept very low, in general it would seem that its effects are beneficial.

DEFICIENCY

Although Chapter 2 has been devoted to scurvy, its emphasis is historical and it would be inappropriate to ignore this classic deficiency disease at this juncture. Scurvy is characterised by a long list of signs and symptoms including bleeding gums, depression, easy bruising, impaired wound and fracture healing, irritability, joint pains, loose and decaying teeth, malaise, and tiredness. Diagnosis may be made on the basis of the more restricted observations of small skin discolorations due to ruptured blood vessels (petechiae) and easily produced bruises (ecchymoses and purpurae), red vascularisation with petechial and lineal haemorrhages on the undersurface of the tongue (sub-lingual haemorrhages), and the hyperkeratosis of hair follicles, with uner-rupted curled hairs and a surrounding pink halo. This latter sign alone is virtually pathognomic of florid scurvy. All these signs appeared within one to two months of an enforced scorbutic diet undertaken by volunteer medical students and after about one month a clinical diagnosis would appear possible on the basis of a low plasma level of 0.2 mg% or less of ascorbic acid (normal average about 1.2 mg% with a range of 0.6–2.5 mg%) and a low leucocyte ascorbate assay of 15–18 μg per 10^8 cells (normal value at least 25 μg per 10^8 cells). A further clinical diagnosis is afforded by the Hess test in which a pressure of 100 mg of mercury from a blood pressure cuff is maintained for five minutes producing characteristic petechial counts. Another haematologic abnormality is anaemia, probably caused by a combination of impeded iron absorption, aberrant folic acid metabolism, and a failure to manufacture red blood cells (erythropoiesis) in the bone marrow; all processes thought to be dependent on vitamin C. After only two weeks on a vitamin C free diet, and while plasma levels were still at a normal value of 1.2 mg%, some volunteers reported psychological disturbances including hypochondriasis and depression. While scurvy sufferers' skin lesions clear up completely on treatment with vitamin C (typically 200 mg per day for a week followed by a lower daily dose) the vascular damage may be enduring, with the sub-lingual varicosities leaving permanent fissures. The more obvious signs of scurvy already mentioned are mainly explicable in terms of a failure to replace the collagen component of damaged connective tissue surrounding all animal cells and forming a major part of bone and dentine. In a recent USA study, induced vitamin depletion in young men caused gingival inflammation and crevicular bleeding. It was implied that the recommended daily dietary allowance of about 60 mg per day was necessary to maintain peridontal health.

It is important for the clinician to recognise the signs and symptoms of scurvy since untreated patients die. Although the view is often expressed that deficiency diseases have been eliminated in the Western World that is not the case, particularly among the urban poor. Vitamin C deficiency is now recognised as common among the chronically sick, the institutionalised elderly, men living alone, those avoiding acidic foods because of digestive problems like hiatus hernia, chronic alcoholics (who are also, and more seriously, likely to be deficient in B vitamins), patients receiving peritoneal dialysis and haemodialysis, and infants fed unsupplemented cows' milk. In the 1970s in the UK it was thought that 80% of the elderly living at home were suffering from long standing vitamin C deficiency and the situation was probably worse in the hospitalised elderly. By contrast, elderly vegetarians were in rude health, ascorbutically speaking, typically with plasma levels of 1.02 mg% and leucocyte concentrations of 36 μg per 10^8 cells. Most UK state school meals are almost certainly deficient and two surveys of young, supposedly fit men employed in the outside building industry and in the Royal Navy showed about a third of them to be exhibiting the skin and sub-lingual signs of the scorbutic. Heavy smoking and drinking, non-salad eating members of the lower social classes probably rely too heavily for their dietary intake on potatoes and in many cases the processing and cooking methods employed reduces this source of vitamin C to insignificance. Other factors that depress tissue levels of ascorbic acid are said to include oral contraceptives and a whole host of diseases and disorders. For a healthy individual it is said that as little as 10 mg per day of the vitamin will prevent the onset of scurvy which apparently manifests itself when the total body pool of ascorbic acid falls below 300 mg.

MAINTENANCE OF HEALTH

Recommended daily dietary allowances (RDAs) over the past few decades have ranged from as little as 10 mg per day to a thousand times as much, depending on ever changing individual, national, and international opinions. Among the nations the USA recommendations are relatively high with the UK levels more in accord with those of the World Health Organisation. Fairly recent RDAs for adults in the UK, USA, and USSR were 30, 60, and 90 mg per day respectively. What is not generally in dispute is that intake needs to be adjusted for age, stress, physiological state, and perhaps sex. Fashions and ideas change with great rapidity but at present the regimes shown in Table 6.1 would attract reasonable approval. The RDAs of the USA are designed for 'the maintenance of good nutrition of practically all healthy people in the USA', and while this appears quite straightforward there can be great difficulty in exactly meeting these criteria in practice because these are total intakes and it is difficult for an individual to gauge the vitamin content of his or her diet.

In reality some people assume that their diet provides all that is needed, and a diet including a reasonable amount of fresh vegetables and fruit does suffice, whilst others ignore their dietary intake and take supplementary vitamin tablets, and the majority probably soldier on in blissful ignorance. Those in the last category run into problems when, through the laziness or ignorance most often expressed by the single elderly, they forsake a balanced diet for convenience foods like pot noodles, tea, and toast. The fastidious pill taker is also faced with difficulties of the opposite type since most vitamin C tablets appear to contain more than 60 mg. Confusion can occur because of the labelling which may read for example, 'Vitamin C, tablets B.P. 100 mg', and, since the stated ingredients are many, the lay person may not be clear if each tablet has a mass of 100 mg or whether each tablet contains 100 mg of the vitamin. The latter is in fact the case. Vitamin C tablets, as with several other types of pills, have been occasionally documented as causing local damage when lodging in the œsaphagus after swallowing whole. This causes a burning chest pain and produces a mucosal ulcer which heals without a scar within a few days. The advice is to be upright and take several mouthfuls of water when swallowing the tablet or, much better, to chew the tablet before swallowing. The

Table 6.1 *Recommended daily dietary allowances (RDA's) for vitamin C*

	RDA/mg per day
Infant	35
Child	45
Adolescent	50
Adult	60
During pregnancy	80
During lactation	100
Old age	150

rather sharp taste of ascorbic acid tends to make the latter method resemble an exercise in masochism but presumably accords with the notion that medicine will only do good if it has an unpleasant taste!

In general there would appear to be no harm in exceeding the RDA though some would say that to do so may be of more benefit to the pharmaceutical industry than the individual, despite the claims of the supporters of the megadose regime (1–40 g per day) as first advocated by the double Nobel laureate Linus Pauling. Indeed it is difficult to see how a very large dose can be of benefit since absorption is limited and, while plasma levels reflect intakes up to about 150 mg per day, the ingestion of larger amounts has little further effect because of a decreasing absorbance (95% of a 100 mg is absorbed but only 20% of 5 g is absorbed as daily doses), the low renal threshold assuring that excess plasma levels are promptly excreted and the rapid increase in inducement of the liver ascorbate catabolising enzymes. When the plasma level reaches 1.4–2.0 mg%, the tissues are probably saturated with a body pool of somewhere between 1500 and 3000 mg and any excess dietary vitamin C is eliminated through the kidneys as ascorbate, dehydroascorbate, ascorbate-2-sulphate, and ethanedioic (oxalic) acid.

A deficiency state may be indicated when the plasma level drops to less than 0.5 mg% and, in general, the plasma level is considered to be a good marker for body stores of the vitamin. The plasma level has been reported to be reduced in many pathological states, *e.g.* infective disorders, congestive heart failure, kidney and liver disease, gastrointestinal disturbances, purpura, endocrine cases, and malignancies. Patients who are febrile, undergoing surgery, or subject to trauma need more vitamin C in their diet, as do smokers, patients receiving parenteral nutrition (to compensate for increased urinary losses), and those exposed to cold and heat stress. It was found that South African mine workers needed approaching 300 mg per day to maintain a serum ascorbate level of 0.7 mg%, although it was suspected they may have been suffering from marginal scurvy before the investigation began. Measuring the plasma concentration has outmoded the 'saturation test' in which a large oral loading dose is given. If the tissues are saturated most of the dose is excreted during the next 24 hours, but if the tissues are deficient then a large part of the dose is retained. Interpreting plasma analysis data may prove difficult if the normal range is as wide as the 0.6–2.5 mg% suggested by many clinical chemists and an alternative strategy is to assay leucocytes, which should normally contain at least 25 μg per 10^8 leucocytes, which may better reflect tissue levels.

For those, undoubtedly the majority even in Western countries, who rely on natural dietary sources of the vitamin, care is needed in culinary practice if much of the ascorbic acid is not to be lost. As seen in Chapter 5 the fine cutting of vegetables releases ascorbic acid oxidase which will destroy the vitamin and the use of excessive water for cooking will leach it out of the food. Overcooking and the addition of sodium bicarbonate, thankfully by now little practised, also destroy the vitamin by oxidation which is particularly rapid in cooking pots made of copper. For the human infant, as for any mammal, mother's milk is a whole food and that includes vitamin C at a level of 3.0–5.5 mg%. Bovine milk is much less rich and needs supplementation for feeding to human infants.

At present there is no certainty that the RDAs represent an optimal intake and the range between the deficient and the toxic is very uncertain. An interesting comparative approach for generating ideas on what may be an optimal rate of intake for man has been to look at rates of synthesis in a range of mammals and to investigate body pools in some species of the more closely related, ascorbate dependent, primates. Several species of ruminants and rodents, the cat, the dog, and the pig synthesise ascorbic acid in the microsomal fraction of their livers at rates which vary from 5 mg per kg body weight per day (the lowest of the ranges for the cat and the dog) to 275 mg per kg body weight per day (the highest of the values recorded for the mouse). The human RDA of 60 mg per day would be equivalent to a synthesis rate of a paltry 0.9 mg per kg body weight per day! The studies with other primates also lend credence to the recommendations for higher intakes, since, despite being able to prevent scurvy in the vervet and rhesus monkeys with a low dose (equivalent to 0.6 mg per kg body weight per day), to maintain good health and/or the body pool of a healthy wild animal feeding rates were needed ranging from 3 to 35 mg per kg body weight per day in vervet, rhesus, squirrel, and macaque monkeys and the baboon. These findings together with the observation that the guinea-pig needs up to ten times its antiscorbutic dose to maintain good health, heal wounds quickly, and recover from anaesthetics all suggest that using scurvy alone to estimate total need may lead to a gross underestimate.

THERAPEUTIC USE

In addition to the treatment of scurvy, now a rare occurrence among the general population, there are many applications, some of sound provenance, others highly controversial,where vitamin C is administered either orally or intravenously, typically in 3 × 100 mg daily doses. This is routine treatment in cases of surgery, physical trauma, and duodenal ulcer where the vitamin will not only aid in wound healing but also in enhancing the body's immune response in preventing and fighting infection. This aspect of the vitamin's physiology is also used in its prescription in infection, fever, diarrhoea, and cases where the risk of infection and inflammation is high, *e.g.* in chronic haemodialysis cases, and it has also been suggested for psychogeriatrics, the institutionalised elderly, pregnant women, and neonates and their mothers. 0.5–3.0 g per day has been given to acidify the urine in cases of refractory urinary tract infections. The vitamin has been described as an immunomodulator and appears to act at various points in the immune system, for example in inhibiting histidine decarboxylase, thus preventing the formation of the immunosuppressive histamine, in enhancing the activity of neutrophil leucocytes, and in neutralising excessive levels of reactive oxidants produced by phagocytes in cases of chronic infection.

Several problems associated with the blood and the circulatory system are treated with vitamin C. Anaemia accompanying scurvy has already been mentioned but general anaemia caused by iron deficiency also responds to vitamin C, although simultaneous iron supplementation is also usually necessary. Ascorbic acid aids in iron absorption by forming soluble iron complexes and reducing iron(III) to iron(II) thus countering the intestinal binding of iron by dietary phytates and tannins. Once the blood iron level is restored it may be maintained by adopting a good iron-containing diet with each meal also containing 25–50 mg of ascorbic acid. It has been suggested that primitive man's diet was rich in meat (providing 35% of his energy) and ascorbic acid (up to 40 mg per day) and that, even if his life was nasty, brutish, and short, it was probably not anaemic!

For haemoglobin to carry oxygen the iron in the haem molecule must be in the reduced iron(II) state. Normally over 98% of haemoglobin is in this form with less than 2% being in the non-functional iron(III) methaemoglobin state. Normally the small amount of methaemoglobin found is reduced to haemoglobin by the enzyme NADH/methaemoglobin reductase (also called erythrocyte cytochrome b_5 reductase). Various congenital methaemoglobinaemias are known and those due to a deficiency in the cytochrome b_5 reductase system respond to treatment with 500 mg per day ascorbic acid or 100–300 mg per day Methylene Blue given orally. It appears that the ascorbic acid directly reduces methaemoglobin, albeit slowly, whereas Methylene Blue activates a normally latent NADPH dehydrogenase by providing the missing link furnished by the cytochrome in the NADH system. This type of methaemoglobinaemia is mild and the treatment is merely to give cosmetic improvement in cyanosis. Another inherited methaemoglobinaemia is due to a defect in the haemoglobin molecule and although this does not respond to ascorbate or Methylene Blue the condition is mild in the extreme. There is some difference in opinion as to the efficiency of the vitamin in treating induced or acquired methaemoglobinaemia as suffered by workers occupationally exposed to various chemicals (*e.g.* trinitrotoluene, nitrobenzene, aniline, naphthoquinone, naphthalene, rescorcinol, and phenylhydrazine) and consumers of various drugs (*e.g.* phenacetin, primaquine, benzocaine, and *p*-aminosalicylic acid). Nitrites also cause the formation of methaemoglobin. Those who oppose the use of ascorbic acid in these cases do so on the grounds of its slow effect compared with Methylene Blue, and in acute cases speed is of the essence otherwise collapse, coma, and death may follow severe cyanosis.

Methaemoglobinaemia is 'naturally' produced by the superoxide radical O_2^- which is normally kept in check by the enzyme superoxide dismutase (SOD) which may also require vitamin C as a co-enzyme. It has been suggested that the administration of vitamin C may reduce the disease severity in sickle cell patients whose red blood cells are low in the vitamin and vulnerable to oxidant damage.

There is some evidence that higher doses of the vitamin increase the circulating levels of lipoproteins (HDL) thereby preventing the arterial deposition of cholesterol and reducing the risk of coronary heart disease. While plasma and leucocyte levels are depressed in coronary heart disease, it is not known which, if either, is the cause and which the effect. However, it has been suggested that supplementation may help to prevent atherosclerosis by maintaining artery-wall integrity (dependent on adequate levels of hydroxyproline essential for collagen synthesis), reducing blood cholesterol (by promoting bile acid synthesis) and in reducing blood tryglycerides (by enhancing the activity of plasma lipase). Other contributions of the vitamin to a healthy circulation may include reducing platelet aggregation and increasing fibrinolytic activity. In experimentally induced scurvy a very few human volunteers have suffered cardiac emergency. Also, arteriosclerosis type lesions have been discovered in rodents and piglets with a chronic deficiency. One authority has christened vitamin C 'the heart vitamin', but although there appears to be a correlation between the incidence of ischaemic heart disease (IHD) and low plasma vitamin C, again, the latter may be a consequence of the former rather than its cause. Nevertheless the risk factor in IHD is thought by some to be various aggressive oxygen species, *e.g.* the superoxide radical, the destruction of which appears to depend on the vitamin C mediated superoxide dismutase.

In the rather common illness idiopathic thrombocytopaenic purpura (ITP), an autoimmune situation exists, which can be induced by certain drugs, in which antibodies are directed against the platelets which are consequently destroyed by the spleen. This results in excessive bleeding and bruising which is normally treated by the administration of adrenocorticosteroids, vinca alkaloids, immunosuppressives, danazol, or i.v. gammaglobulin or, more effectively, by splenectomy. However, all the treatments have severe side effects and are by no means 100% effective. In 1988, one refractory patient took vitamin C and her platelet count rose, which resulted in her physicians organising a trial during which seven out of eleven refractory patients responded to a daily dose of 2 g of vitamin C with increased platelet counts and intravascular platelet survival times and a decrease in bruising and petechiae. This non-toxic treatment has been repeated in other hospitals with great success with some patients but as one specialist in a Belfast hospital told me 'We don't know how it works!'

A much rarer disease is an inherited defect in the production of lysosomes leading to clinical features due to a resulting malfunction of melanocytes (albinism), platelets (mild bleeding disorders), and phagocytes (lowered resistance to bacterial infection). This is known as Chédiak-Higashi disease and is usually fatal in the first or second decade due to uncontrolled infection, but some patients have had their microcidal defect corrected by large doses of ascorbic acid (20 mg per kg body weight per day).

Various inherited defects in the synthesis of the α- and β-chains of haemoglobin result in a family of anaemia diseases known as the thalassaemias. In the more severe cases, such as β-thalassaemia major, the only life sustaining treatment is regular transfusion resulting in a gross iron overload as an inevitable side effect. This is countered to some extent by infusing the drug desferrioxamine (deferoxamine, desferral), which increases the urinary output of iron, and the effect can be enhanced by the simultaneous administration (normally oral) of ascorbic acid in doses that have ranged from 150 mg to 2 g per day. The use of the vitamin in thalassaemia treatment is now the matter of some controversy and is discussed later in this chapter under 'Toxic Effects'.

The vitamin has also been routinely prescribed in cases of threatened abortion, thyroxidosis, achlorhydia, diarrhoea, prickly heat, rheumatic fever, rheumatic arthritis, and in cases of spinal injury to reduce urine acidity. The physiological basis of these therapeutic applications is not entirely clear except in the cases of achlorhydia and diarrhoea where there is a risk of anaemia caused by a reduction in the intestinal absorption of non-haem iron which is enhanced by vitamin C.

Recent research and epidemiological studies have suggested that low vitamin C status is associated with cataract and increased intraocular pressure, diabetes, smoking, and moderate to heavy alcohol consumption. Cataract is common in old age, when vitamin C status is often low, and particularly in men. One cause of cataract in the elderly is thought to be oxygen and its radicals, particularly the superoxide ion O_2^-, which, as we have seen is normally destroyed by the ascorbate dependent superoxide dismutase (SOD) enzyme. Experimentally induced *in vitro* photochemical damage in animal lenses is reduced by ascorbate and human lenses affected by cataract have been found to be significantly lower than normal in vitamin C concentrations. In guinea-pigs dosed with vitamin C, the ultraviolet light-induced damage to a lens exopeptidase (thought to prevent the accumulation of cataract-forming proteins) was inversely proportional to the dose administered. In a trial involving 450 patients with incipient cataract, a dose of 1 g of vitamin C per day reduced full cataract development to a level less than expected. The investigations on intraocular pressure suggest that the elderly may benefit from an intake of more than 150 mg per day, although the success with quite large doses given to glaucoma patients has been somewhat mixed. The topical application of a 10% aqueous solution, given three times a day, was found to be beneficial with one group suffering from glaucoma.

The recent Australian discovery that diabetics can have body levels of vitamin C 70–80% less than normal supports previous work carried out in India and it has been suggested that it may be at the root of complications like heart and kidney failure, blindness, and gangrene. One hypothesis suggests that chronic hyperglycaemia may be associated with intracellular deficits of leucocyte ascorbic acid, on the basis that glucose and ascorbic acid are sufficiently similar to share or compete for the same membrane transport system, leading to the impairment in the acute inflammatory response, increased susceptibility to infection and faulty wound-repair characteristic of the untreated diabetic. It is not clear whether diabetics absorb less or excrete more of the vitamin than normal but the suggestion is that they should benefit from supplementation which may improve glucose tolerance. Very large doses, however, should be avoided as this leads to an increase in blood levels of dehydroascorbic acid, which has been shown to produce diabetes mellitus in rats!

The well known toxic effects of alcohol appear to be ameliorated by the administration of vitamin C which plays a role in hepatic detoxification mechanisms by contributing to the oxidising cytochrome P450 system. Perhaps the gin and tonic imbibers should be advised to resume the now outmoded gin and orange in order that the latter may help to counter the after effects of the former! The vitamin has also been used in treating lead poisoning but carefully controlled trials showed no effect of ascorbate in reducing heavy metals in hair.

It has been suggested that the involvement of vitamin C in converting the essential fatty acid arachidonic acid to prostaglandins is implicated in maintaining the tone and reactivity of the airways. In this context the vitamin may well be of benefit in alleviating the distress of those suffering from asthma and emphysaema. A dose of 1 g daily has been claimed to prevent asthma attacks in some patients and in others a single dose in the range 4–8 g has had an apparently ameliorating effect during an attack. Emphysaema and bronchial carcinomas may owe their origin to an increase in polymorphonucleocytes and macrophages (stimulated, for example, by cigarette smoke) causing an increase in reactive oxidants. As has already been mentioned vitamin C, among others, is an important agent in neutralising reactive oxidants. Smoking reduces serum levels of the vitamin by up to 0.2 mg% and it has been suggested that the smoker may need an additional 60–70 mg per day to compensate for this effect. It is not clear if the low level of plasma ascorbate of smokers is due to an enhanced rate of metabolism, a reduced absorption, or a reduced intake due to their habit being indulged to the exclusion of the casual eating of fruit.

Vitamin C has also been recommended in the treatment or prevention of the common cold, mental illness, infertility, cancer, and AIDS. These treatments have aroused great controversy and are more or less discountenanced by the mainstream of medical opinion.

Despite controlled trials proving nothing conclusive regarding the common cold, many individuals are sure that regimes ranging from an orange a day to daily megadoses of several grams of vitamin C either prevent or alleviate this all-too-common infection. At best, some trials indicate a possibly significant reduction in the number of days of disability. There is some evidence that vitamin C stimulates the thyroid gland and may therefore raise the metabolic rate which, in conjunction with the vitamin's undoubted antihistaminic and immunological role, gives some support to the notion that the vitamin encourages a sense of good health. This type of argument has also been used in support of its claimed role in longevity. Many would subscribe to the view that if you feel well you are well but the slogan that vitamin C represents 'youth in a bottle' seems to be going too far. What is clear is that plasma and leucocyte levels of the vitamin become very depressed at the onset of a cold or any other infection. This is probably due to ascorbate being oxidised as a consequence of the increased activity of the neutrophils but there is no conclusive proof that dietary supplementation will prevent such infections.

Although, in controlled trials, no effects on male fertility could be demonstrated, ascorbic acid has been prescribed at 1 g per day for ten days followed by 500 mg per day through two of the partners' menstrual cycles where infertility was suspected of having been due to sperm agglutination. Agglutination is probably caused by the undesirable oxidation of a protein present in seminal fluid. The reduction of this protein may be restored by vitamin C, in the presence of which it is said that sperms swim singly. Human seminal fluid is normally high in ascorbic acid (about 12 mg%) compared with serum (for example 1.2 mg%).

Claims that very large doses of the vitamin have restorative effects in cases of Parkinsonian symptoms, schizophrenia, depression, Alzheimer's disease, or the IQ of the mentally retarded have been described by one authority as, at best, a proposition of very doubtful validity. Nevertheless there is good evidence that certain psychiatric patients may be in a state of sub-scurvy and in such cases supplementary vitamin C may be beneficial. There is evidence that some depressive states may be caused by pentavalent vanadium and vitamin C may protect by reducing it to the harmless quadrivalent form. Many successes have been claimed for moderate oral doses in reducing mania, skipping long established depressive periods and reducing tantrums in children with learning disabilities, but these are by way of anecdotal incidents rather than the results from controlled double blind trials. The reasoning of the vitamin's champions in the context of mental illness is probably based on its well accepted biochemical role in the synthesis of various neurotransmitters as described in Chapter 5 and there is now a revival of interest in a possible connection between vitamin C deficiency and Alzheimer's and senile dementia.

The most heatedly controversial area of vitamin C therapy concerns the treatment or prevention of cancer. The Nobel prize-winning biochemist Linus Pauling has been one of the most active supporters of the megadose regime and, in 1973, together with Scottish surgeon Ewan Cameron, he published results which demonstrated a remission effect of a better and longer life for cancer patients given up to 10 g per day. Suffice it to say that the medical profession in general are more sympathetic to other researchers who cast doubt on the reliability of Pauling and Cameron's data and/or experimental design and were unable to repeat the effects in major trials carried out in the USA. There is, nevertheless, a sneaking feeling that the vitamin may enhance the cytotoxicity of several chemotherapeutic agents. More recent large-scale trials carried out in Europe suggest that the vitamin may well protect against the onset of gastric cancer and the effect may be related to the vitamin's demonstrated *in vitro* inhibition of nitrosamine formation. Nitrosamines may be formed as a result of an interaction between nitrites and dietary amines and are thought to be a significant cause of gastric and oesophageal cancer. Although the ingestion of nitrite is generally small it can be produced by the reduction of nitrate by bacteria in the gut and this is the reason for the widespread concern regarding the rise of nitrate levels in drinking water. The geographical and cultural correlation of gastric cancer are strongly indicative of environmental or dietary effects with the risk of stomach cancer being related to a deficiency of fruit and vegetables containing vitamin C. The incidence of gastric cancer in the USA, while never high, has nevertheless fallen significantly since the introduction of regular vitamin C supplementation. From the results of other recent trials it has been inferred that the vitamin may protect against the initiation of cancer of the cervix. Recently, a platinum derivative of ascorbic acid has been shown to cure certain types of mouse tumours. The administration of megadoses of the vitamin to colonic cancer patients following chemotheraphy and radiation treatment has produced no measurable benefit, however, and the notion that the vitamin could replace established surgical, chemotheraphy, and radiation practices has been strongly refuted. Despite these conflicting views some authorities suggest that human populations that are deemed as high risk with regard to developing cancer could benefit from high daily doses of 1–2 g per day. This seems to be effectively saying that it probably will do no harm and there is just a chance that it might do some good!

Linus Pauling has re-entered the fray recently by supporting research on vitamin C as a possible treatment for Aids patients. The Californian homosexual community is known to believe that massive vitamin supplementation in the range 50–200 g per day is efficacious both in preventing and improving, if not curing, the condition. Pauling has said that, although this may be wishful thinking, investigation is needed, if only to lay to rest false hopes.

In a recent publication dealing with nutritional influences on illness, it is said that vitamin C supplementation may be beneficial in preventing or curing no less than forty disease states. In addition to those already covered are included bursitis and vascular fragility (commonly encountered in sports injuries), Crohn's disease, cholesterol gallstone formation, gout, hepatitis, *Herpes simplex* infections, menorrhagia, multiple sclerosis, obesity, osteoarthritis, rheumatoid arthritis, tiredness, duodenal and gastric ulcers, bed sores, chronic urticaria, and wound healing. In an impressive trial with post-surgery patients, where serum hepatitis is a highly undesirable after-effect risk, of the 1095 patients given 2 g of vitamin per day, none succumbed, whereas nearly 10% of the 150 patients given less than 1.5 g per day did. Massive doses of up to tens of grams appeared to hasten the recovery of patients with viral hepatitis. Vitamin C, in combination with bioflavanoids, reduced the blistering and halved the duration of outbreaks in individuals just starting the signs of a *Herpes simplex* incident. The combination of 250 g of vitamin C with 100 g of zinc sulphate, taken twice a day by recurrent herpes sufferers, gave 100% suppression of outbreaks, with symptoms being restricted to a mild tingling in many cases. Another long established use of vitamin C and zinc, given in many hospitals, has been in the treatment of decubitus ulcers (bed sores) which seem to respond to both oral and topical applications. In most therapies the vitamin is given orally but intravenous injections of 2 g doses given three times a week for six or seven weeks improved healing rates in a majority of patients suffering from duodenal or gastric ulcers. In experiments with mice, the severity of heat burns and frostbite was reduced in animals given an ascorbic acid supplement although the effect on human burns is not known.

TOXIC EFFECTS

Out of the heat generated from the debate started by Linus Pauling's recommendation of megadose prophylaxis the notion that large doses of vitamin C may *cause* cancer emerged. Indeed an article in the *New Scientist* in 1985 was entitled 'Vitamin C: Does C stand for Cancer?'! The evidence for this view is based mainly on animal experiments and it is difficult to extrapolate this to the human situation although concern has been expressed regarding a rebound effect in those who may suddenly *cease* a high dose which leads to circulating levels of the vitamin dropping well below normal. This, albeit temporary, effect could make a person more vulnerable to the development of undesirable diseases including tumour development. Any sudden discontinuation of even a moderate level of supplementation could cause rebound scurvy and it is advised that, if supplementation is to be stopped, it should be tapered over days or even weeks.

Another concern is the fear that, since the vitamin is metabolised mainly via ethanedioic (oxalic acid), a high dose might lead to the formation of kidney stones. While there is no clear evidence that this occurs, and the phenomenon has been described as a myth, the commonsense advice is for those prone to recurrent stone formation or suffering from any renal impairment to avoid high doses which also exacerbate acidosis in chronic renal disease and renal tubular acidosis. Sustained hyperoxaluria has been found only rarely to accompany an excessive ingestion of vitamin C.

In some cases megadoses in the range 1–2 g per day have produced mild ill effects of the intestinal tract including discomfort, pain, and osmosis diarrhoea. Realistically diarrhoea is almost certainly the only major side effect of over large doses for the otherwise healthy.

Dietary levels in excess of 600 mg per day were found to decrease the oxidase activity of human serum ceruloplasmin. It was suggested that this may be due to dissociation of the copper co-factor but no suggestion is made as to whether this may lead to illness like Wilson's disease in which ceruloplasmin levels are reduced and excess copper is deposited in the tissues. Very recent work has shown an enhanced uptake of copper from ceruloplasmin by human cells treated with both L-ascorbate and its D-isomer.

Reference has already been made to the widespread practice of administering vitamin C to enhance the effect of drugs that increase the urinary output in patients suffering from iron-overload (particularly those suffering from thalassaemia and congenital sideroblastic anaemia). For some time, there has been concern that, under these circumstances, the vitamin increases the toxicity of iron and there is evidence of a deterioration of left ventricular function and a general aggravation of haemochromatotic myocardiopathy in some patients. The current view is that the use of vitamin C supplementation to enhance responsiveness to deferoxamine therapy should be carried out with extreme caution, if at all, particularly in the case of older patients. From animal experiments, it appears that, at the cellular level, damage occurs by peroxidation of membrane lipids leading to injury to the heart muscle and a reduction in phagocytic activity resulting in increased infection. The undesirable action of ascorbate may be to reduce the deferoxamine liberated iron(III) to iron(II) which triggers the so-called Fenton reaction in which iron(II) reacts with endogenous hydrogen peroxide to release the highly reactive and specially dangerous hydroxyl radical OH·. The hydroxyl radical inflicts membrane injury by causing lipid peroxidation. Lipid peroxidation is accompanied by the release of malonyldialdehyde, which has been used as an indicator of membrane damage and is found to be depressed by the administration of deferoxamine but increased by the co-administration of ascorbate. Interestingly, the Bantu tribesmen of South Africa, who traditionally ferment beer in iron vessels and subsequently carry a high loading of body iron, appear not to suffer from iron toxicity, perhaps because of their low vitamin C status.

Some other causes for concern are those individuals who are deficient in glucose-6-phosphate dehydrogenase, who may be at increased risk of a haemolytic episode following a megadose, and infants born from mothers taking a regular megadose, who may develop scurvy, having become dependent on a high foetal regime.

In addition to directly toxic effects, medical mishaps could arise through high doses causing interference with the clinical analysis of blood and urine for various metabolites. False negative and false positive results for different types of urine glucose tests can occur and a false negative glucose test can obscure the presence of occult bleeding. Automated procedures for measuring alanine amino transferase (ALT) or glutamate pyruvate transaminase (GPT), lactate dehydrogenase (LDH), and uric acid in body fluids can also be affected when patients are receiving in excess of 3 g per day.

VETERINARY USE

Intensive methods of keeping livestock causes stress which can impair performance in, for example, the growth of young animals, the laying rate of hens, and the shell strength of their eggs. Trials have demonstrated that such stress induced problems can be reduced by giving vitamin C even though the animals concerned are capable of synthesising ascorbic acid. Pre-ruminant calves, however, have a very limited synthetic ability and are naturally dependent on milk. Calves that are rapidly weaned from suckling, as is the practice in intensive modern agriculture, need a vitamin C supplement in their weaner feed. It is found that occasionally dogs suffer from a disease strongly resembling scurvy. This is likely to be do to a congenital failure of the dog to synthesise its own ascorbic acid and such animals benefit from the administration of vitamin C in their feed. Much work has been carried out on ascorbic acid in fish. It is found that an insufficient supply of the vitamin results in deformations of the vertebrae column, a tendency towards infection, and poor wound healing. The ascorbic acid status of fish is determined from the concentration in the liver, the hydroxyproline:proline ratio in the skin and the collagen content of the vertebrae.

Although a definitive role for vitamin C, other than being involved in producing the hydroxylated amino acids essential for connective tissue formation, awaits discovery, it seems certain that many kinds of ill-health other than scurvy are kept at bay by this simple, bitter-sweet substance. There is little agreement on what constitutes an optimum ideal daily intake but it seems to be in the range 50–500 mg, depending on state of health, age, and social habits. While a balanced diet containing fruit and vegetables should provide an adequate intake it is likely that an increasing proportion of the population needs supplementation in the form of tablets which are typically formulated with lactose, glucose, tartaric acid, maize starch, stearic acid, and magnesium stearate. Most of these substances are necessary in binding the ingredients to produce a strong tablet, which commonly contains 100 or 500 mg of ascorbic acid. Rather more pleasant than these basic tablets, but more expensive and usually containing 0.5 or 1.0 g of ascorbic acid, are the chewable and soluble effervescent tablets that are fruit flavoured, sweetened and, in the latter case, laced with citric acid and carbonates to produce the fizz.

As has been remarked elsewhere, some higher authority ordered the deletion of the gene many millions of years ago which makes man a non-synthesiser, unlike the majority of other mammals. This dietary dependence seems an unfortunate inconvenience but it has been argued that this genetic shortcoming, like sickle cell anaemia, may confer some benefit. This line of argument, however, may be more a case of being determined to make a virtue from a necessity.

Chapter 7

Inorganic and Analytical Aspects of Vitamin C Chemistry

L-Ascorbic acid is now produced in thousands of tonnes every year. It is used very extensively in the food industry and has its own E number (E300). Many foods have it added simply as a vitamin supplement, *e.g.* in fruit juices. In bread making it forms part of the baking process and is used as an antioxidant in a wide variety of foods. It is sold as an 'over-the-counter' medicine in the form of pills and as a component of various multivitamin tablets; indeed in shops in California it may be bought in 1 kg containers in powder form. It is clear, therefore, that as part of the quality control and assurance procedures it is necessary to have reliable and accurate analytical procedures. Much remains to be discovered about the role of vitamin C in living systems and as part of such studies analysis of very small quantities of the vitamin in many different matrices derived from both plants and animals will be required.

It might at first glance appear strange to attribute an inorganic chemistry to an organic molecule such as L-ascorbic acid. However, the very complex nature of the redox chemistry of the molecule makes it both an interesting and intriguing reducing agent in inorganic systems. This is why much effort has been expended on the investigation of the reactions of L-ascorbic acid with metal ions and metal ion complexes, which are also involved in the biochemistry of vitamin C. The enzyme ascorbate oxidase for example has copper atoms incorporated into its structure and these undoubtedly play a vital part in the operation of the enzyme.

ANALYTICAL CHEMISTRY

The analysis of L-ascorbic acid in its various forms presents many difficulties and even today there is no universal, routine method of analysis which is free from interferences. Any procedure should be capable of distinguishing between L-ascorbic acid and its various oxidation products. It should also be capable of assaying ascorbic acid and these oxidation products simultaneously with a minimum of sample preparation and interference from other species present. This is a tall order. The situation is complicated by the fact that vitamin C occurs in a very wide variety of plant and animal tissues along with a host of other organic molecules from which it has to be distinguished either by separation or by measuring some unique property of L-ascorbic acid. Perhaps the most obvious property of L-ascorbic acid which could be used for assay is its redox behaviour and this has been a feature of many analytical procedures. Other methods have involved the determination of total vitamin C in both the oxidised and reduced forms. This latter is sometimes an advantage since many of the useful medicinal properties of this compound are exhibited by both the oxidised and reduced forms.

Bioassays

These are now of largely historical interest, but they do have the advantage that they are a measure of a particular medical feature, antiscorbutic activity, which is also exhibited by dehydroascorbic acid. The procedures are expensive and very time consuming and give very variable and sometimes unreliable results. Rats cannot be used for the bioassay, since they are able to synthesise their own vitamin C; therefore guinea-pigs are used. The animals are fed with a diet supplemented with various amounts of vitamin C. The guinea-pigs are then killed and their teeth are examined using histological techniques. This allows an estimate of the degree of protection against scurvy afforded by the amount of vitamin C in the diet. An International Unit of vitamin C has been adopted as the antiscorbutic activity of 0.05 mg of ascorbic acid.

Titrimetric and Colorimetric Methods of Analysis

Titrimetric methods take advantage of the redox properties of L-ascorbic acid. Usually, oxidising agents are used to oxidise it to dehydroascorbic acid. A technique which was first introduced as long ago as 1927 utilises the compound 2,6-dichlorophenolindophenol, or 2,6-dichloro-4-[(4-hydroxyphenyl)imino]cyclohexa-3,5-dien-1-one, which is blue at neutral pH and pink in acid solution and which

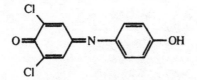

Figure 7.1 *The structure of 2,6-dichlorophenolindophenol*

gives a colourless product on reaction with L-ascorbic acid. The structure is shown in Figure 7.1

When this reaction is used with L-ascorbic acid as the reductant, it is perhaps the most popular titrimetric method for vitamin C. It is very simple to carry out, with a fairly easily detected endpoint, and can readily be used when there is a fairly high concentration of vitamin C and when it is the only redox component in the solution. However, the technique is unfortunately very susceptible to interference from reducing agents. Vitamin C is often to be found in solutions which also contain other reducing agents, notably sulphur dioxide, tannins, metal ions, reducing sugars, *etc*. Methods are available to diminish the effects of these interferences in individual cases, but there is no technique available which allows the removal of the effects of all reducing species in the analyte solution at the same time. Any colour change is of course masked if the solution is coloured and a variety of instrumental techniques such as polarography have been used to detect the endpoint.

Many other titrating reagents have been used or described such as iron(III) when the endpoint is detected using ferrozine, a,a'-dipyridine, or 2,4,6,-tripyridyl-*s*-triazine as indicator (Figure 7.2).

In many redox reactions of L-ascorbic acid which have been studied it is assumed that dehydroascorbic acid is the only product. This is often not the case and the oxidation proceeds further. A technique for unambiguously determining the concentration of L-ascorbic acid, dehydroascorbic acid, and some or all of the further oxidation products with precision, accuracy, and sensitivity in the same solution still eludes us.

Spectrophotometric Analysis

An aqueous solution of L-ascorbic acid is colourless and does not absorb significantly in the visible, although in neutral solutions it has a strong absorption at 265 nm. This would be very suitable for direct

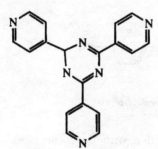

2,4,6-tripyridyl-s-triazine

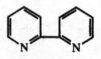

α,α'-dipyridine

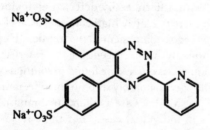

Ferrozine: 3-(2-pyridyl)-5,6-bis(4-phenylsulphonic acid)-1,2,4-triazine disodium salt

Figure 7.2 *Indicators for the titration of L-ascorbic acid by iron(III)*

spectrophotometric analysis, but vitamin C is more often than not found in solutions containing substances which also absorb strongly in this region of the ultraviolet and this severely limits the use of direct spectrophotometry. There has been doubt for some years about the exact molar absorptivity at 265 nm and values between 7500 and 16650 have been given by various workers. The reason for these differences lies in the rapid atmospheric oxidation of L-ascorbic acid in neutral or slightly acid solution. Unless the u.v. absorbance is measured under rigorously anaerobic conditions, a low value will be inevitable since the oxidation products do not absorb significantly at 265 nm. The difficulties are exacerbated by the fact that copper(II) and other transition metal ions are potent catalysts of the oxidation by dioxygen. These have to be removed or complexed by the addition of a chelating agent such as EDTA. The position of maximum absorbance is pH-dependent and drops to 245 nm in acid solution and absorption

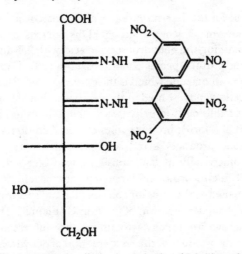

Figure 7.3 *The 2,4-dinitrophenylhydrazine derivative of 2,3-diketogulonic acid*

at this wavelength in a hydrochloric acid, potassium chloride solution has been used to determine vitamin C in soft drinks and in some drugs where there is little interference from other substances present.

It is frequently desirable to analyse for dehydroascorbic acid and L-ascorbic acid in the same solution. The former has an u.v. absorbance at 220 nm, but with a much smaller molar absorbtivity of 720. Thus any attempt at a direct spectrophotometric determination will be factor of about 20 times less sensitive than for L-ascorbic acid. We will see later that these facts have far reaching consequences in the use of chromatographic techniques for the separation and subsequent detection of L-ascorbic acid and dehydroascorbic acid.

In order to attempt to overcome the problem of interferences in the u.v. region from substances found in plant and animal tissue containing vitamin C, reagents have been sought which have specific colour reactions with L-ascorbic acid and/or its oxidation products. The titrimetric method using dichlorophenolindophenol described above has been adapted for colorimetric determination. A coloured 2,4-dinitrophenylhydrazine derivative is also formed with the vitamin which can be utilised for spectrophotometric determination. It is found that the same product is formed with dehydroascorbic acid and with 2,3-diketogulonic acid, oxidation products of L-ascorbic acid. In fact the product is actually a derivative of the last of these, 2,3-diketogulonic acid (Figure 7.3). In practice the reaction is widely used as a technique for the determination of dehydroascorbic acid, which is frequently known as Roe's method. This consists of reacting the solution containing the dehydroascorbic acid with 2,4-dinitrophenyl-hydrazine under quite specific conditions, usually 37 °C for four hours. This produces an osazone which is the derivative of 2,3-diketogulonic acid. The method may be used in the determination of L-ascorbic acid itself by first oxidising it to dehydroascorbic acid using, for example, active charcoal (Norit), bromine solution, *etc.* Addition of strong acid to the osazone produces a red solution which may be measured spectrophotometrically at 530 nm. The deep blue colour which is produced when L-ascorbic acid is reacted with diazonium compounds has also been used as the basis for colorimetric and spectrophotometric methods for the determination of vitamin C (Figure 7.4).

An alternative approach is to utilise the fluorescence which is exhibited by the product of the condensation of dehydroascorbic acid with *o*-phenylenediamine. The quinoxaline produced fluoresces at 427 nm when irradiated at 350 nm. Generally the procedure consists of oxidising any L-ascorbic acid present to dehydroascorbic acid, so that the total may be determined spectrofluorimetrically.

Electrochemical Methods

Electrochemical methods of analysis hold out the possibility of highly selective methods of analysis which offer high accuracy and precision

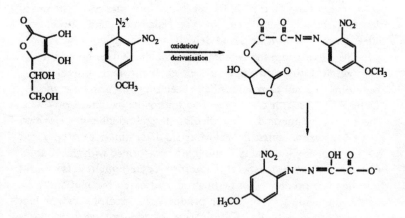

Figure 7.4 *Reaction producing blue diazonium compound with L-ascorbic acid*

as well as great ease of operation. Many methods have been devised. Vitamin C in multivitamin tablets containing iron(II) has been analysed using differential pulse voltammetry at the glassy carbon electrode. However, much more development work is required and these techniques have not been widely adopted.

Chromatographic Techniques

These present the best hope of overcoming the major problem of the effects of many interfering substances which are frequently present in mixtures containing vitamin C. We have seen that there are as yet no techniques which unequivocally allow the direct determination of small amounts of L-ascorbic acid and/or its oxidation products simultaneously in the presence of any other substances. The only satisfactory way of achieving this degree of specificity is to separate the analytes from each other and from other substances present by chromatography.

Until comparatively recently, the most commonly used chromatographic technique in general use was gas–liquid chromatography (GLC). It is not possible to determine L-ascorbic acid directly by this method, because of its involatility, and a time-consuming work-up procedure involving conversion of L-ascorbic acid into its volatile trimethylsilyl ether (Figure 7.5) has been devised to allow the latter to be separated by GLC. This method is capable of delivering accurate and reproducible results, but because of the lengthy preparation required it is not as convenient as the more recently devised high performance liquid chromatography (HPLC).

For some time now, HPLC has often been the method of choice for the rapid determination of a wide variety of organic and inorganic compounds. It is frequently used in the pharmaceutical industry in the analysis of analgesics and other drugs. It has also long been used in the analysis for vitamins. It is however not yet the solution to the determination of vitamin C. Many problems remain and it cannot be

Figure 7.5 *The volatile trimethylsilyl derivative of L-ascorbic acid*

used in all circumstances with equal ease for a variety of matrices and for very small concentrations. However, when it is possible to use HPLC the method is usually reliable, accurate, and reproducible.

The problem of detection of the separated species in HPLC experiments requires a physical property specific to vitamin C (preferably both for L-ascorbic acid and its oxidation products). At the time of writing the most commonly used detectors operate by the measurement of the absorption of u.v. or visible radiation. These allow the detection of nanogram quantities of L-ascorbic acid, but they are much less sensitive to dehydroascorbic acid because of its much lower molar absorbtivity (see above). It remains to be seen whether other detectors such as those which are mass sensitive will produce more effective methods of analysis.

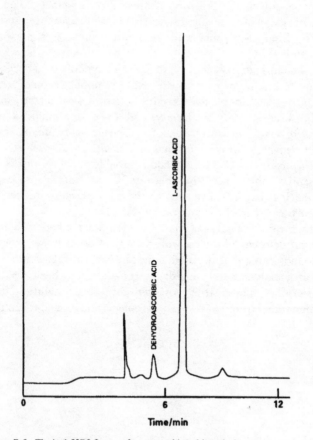

Figure 7.6 *Typical HPLC trace for L-ascorbic acid analysis*

A typical HPLC method involves the use of a column containing a high surface area, small pore size polymeric reversed-phase adsorbent, such as PLRS-s 100A in a low pH aqueous mobile phase. A solution of 0.2M NaH_2PO_4, which gives a pH of about 2.1, is suitable and this also suppresses the action of metal ions as well as giving a well buffered mobile phase. This method has been used in the determination of both L-ascorbic acid and its oxidation products in fruit juices. An u.v. spectrophotometric detector operating at 220 nm (at which wavelength both species have a similar molar absorbtivity) gave satisfactory sensitivity for such media. A typical trace is given in Figure 7.6. It will be seen from this that not only are L-ascorbic acid and dehydroascorbic acid separated by this method, but also peaks due to further oxidation products are present. This is one of the many HPLC methods available for this analysis.

INORGANIC CHEMISTRY

This section will be largely concerned with the reactions of L-ascorbic acid with metal ions and metal ion complexes. However, it would be inappropriate to discuss this subject without a consideration of the properties of the compound which impinge upon inorganic chemistry. It is therefore important to include in this section a discussion of the rather complex redox chemistry of vitamin C and some information on the detection and properties of the species which are frequently cited as intermediates in many redox reactions, namely the ascorbate radicals.

Ascorbic Acid Free Radicals

Since L-ascorbic acid is potentially a two-electron reducing agent it is reasonable to expect that during the course of any redox reaction electron loss from the L-ascorbic acid occurs in two discrete stages. If this is the case, then the intermediate formed from the loss of one electron is some form of ascorbic acid radical. The existence of the free radical was first demonstrated over thirty years ago. Yamazaki and co-workers showed the presence of the radical *in vitro* using electron spin resonance (e.s.r.) spectroscopy. This is a technique which measures the interaction between an incident magnetic field and the spin of an unpaired electron in a molecule. The ascorbate radical has a single unpaired electron and has been observed *in vivo* by the same technique.

The ascorbate radical has been generated in aqueous solution in order to study its u.v. absorption spectrum using the technique of pulse radiolysis. A dilute solution (5 mM) of L-ascorbic acid is bombarded by pulses of 1.9 MeV electrons each lasting about 25 μs. These are typically delivered by a Van der Graaf generator. The solution is saturated with nitrous oxide (N_2O) and this converts the solvated electrons produced into OH radicals. The L-ascorbic acid under these conditions is oxidised very rapidly by the OH radicals to ascorbic acid radicals. This experiment is repeated many times and the absorbance of the solution is measured over a range of wavelengths between 320 and 420 nm at each pulse. The spectrum of the radical may then be built up over this wavelength range. There is found to be a maximum absorbance at 360 nm with a molar absorbtivity of 3700. From these experiments, it is also possible to measure the kinetics of the decay of the radical. The u.v. spectrum of the ascorbic acid radical measured at a pH of 3.3 is shown in Figure 7.7. This has been determined all the way through the pH range -0.3 to 11 and there is no appreciable change in the position of the maximum absorbance in that range, showing that a single species, almost certainly $A^{\cdot-}$, is present throughout this range of acidity.

It is possible to obtain structural information about the radical by the interpretation of the esr spectrum shown in Figure 7.8. It can be seen to consist of a doublet of triplets and each band in the triplets is split into two. This spectrum is fully explained by the structure shown in Figure 7.9.

The presence of only three resonances in the spectrum suggests that the unpaired electron density is spread over all three carbonyl groups.

The decay of $A^{\cdot-}$ is complicated. A reaction scheme which has been suggested involves the combination of two ascorbate radical ions together to form an intermediate which then decomposes to form the products. It is necessary to propose such a mechanism in order to explain the dependence of the rate of the reaction on the ionic strength and the pH of the solution. The mechanism proposed is shown in equations (7)–(9).

$$2A^{\cdot-} \rightleftharpoons (A^-)_2 \tag{7}$$
$$(A^-)_2 + H^+ \rightarrow HA^- + A \tag{8}$$
$$(A^-)_2 + H_2O \rightarrow HA^- + A + OH^- \tag{9}$$

Such ascorbate free radicals are important intermediates in a wide variety of *in vitro* reactions involving oxidation and reduction and there is evidence that ascorbate radicals also have an important role in living systems. Vitamin C is known to interact with the tocopheroxyl radical

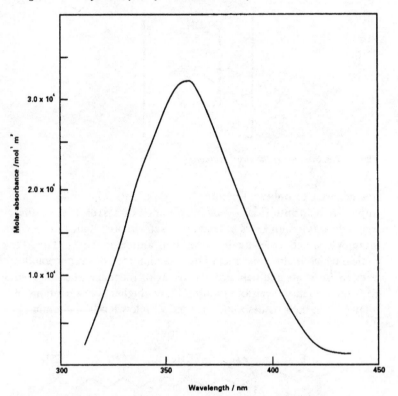

Figure 7.7 *Ultraviolet spectrum of the ascorbic acid radical at pH 3.3*

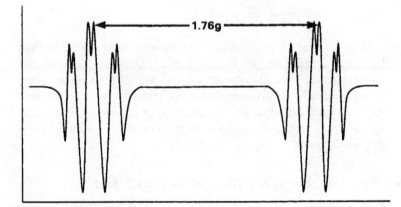

Figure 7.8 *Electron spin resonance spectrum of the ascorbic acid radical*

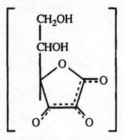

Figure 7.9 *Structure of the ascorbate radical*

to regenerate tocopherol (vitamin E, Figure 7.10). This vitamin is an important lipid antioxidant in biomembranes. In this role the resulting free radical (vitamin E·) is believed to react with ascorbate to re-form vitamin E, which is then available for more antioxidant duty. The A·⁻ radical which is also a product of this reaction then disproportionates back to ascorbate and dehydroascorbic acid. Indeed, guinea-pigs fed with high dietary levels of vitamin C have higher concentrations of vitamin E in their tissues than those fed with low levels of vitamin C.

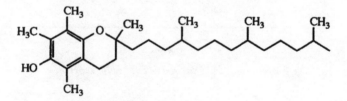

Figure 7.10 *α-tocopherol, vitamin E*

It is believed that ascorbate free radicals have an important biological function in the reaction with destructive free radicals, particularly those associated with oxygen which often cause problems in living cells. They have been shown to quench hydroxyl radical and singlet oxygen, a function that is thought to be particularly important in the eye where comparatively large amounts of vitamin C are frequently found.

Redox Equilibria of L-Ascorbic Acid

In aqueous solution the oxidation of L-ascorbic acid to dehydroascorbic acid appears to be a straightforward example of a two-electron

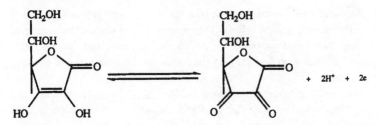

Figure 7.11 *Equation of the oxidation of L-ascorbic acid found in many texts*

redox process. Indeed in many texts the redox reaction is given as that shown in Figure 7.11, with a standard reduction potential of -0.058 V. However, the system is much more complex than it at first appears. The most obvious complicating feature is the fact that any redox process involving L-ascorbic acid is affected by proton transfer, since L-ascorbic acid, its free radical formed from loss of one electron, and dehydroascorbic acid all have acid–base properties. The acid–base properties of these species can be summarised in equations (10)–(13).

$$H_2A \rightleftharpoons H^+ + HA^- \quad pK_a = 4.25 \tag{10}$$
$$HA^- \rightleftharpoons H^+ + A^{2-} \quad pK_a = 11.79 \tag{11}$$
$$HA^\cdot \rightleftharpoons H^+ + A^- \quad pK_a = -0.45 \tag{12}$$
$$A \rightleftharpoons H^+ + A'^- \quad pK_a = 8.0 \tag{13}$$
Conjugate base of A

In addition to these equilibria, in aqueous solution, dehydroascorbic acid rapidly forms a hemiacetal structure.

We can see from these equilibria that L-ascorbic acid is a moderately weak acid in the loss of the first hydrogen ion and a very weak acid in the loss of the second. The pK_a for dehydroascorbic acid is only a rough value. The radical, $HA^\cdot$, however, is a strong acid, comparable with the mineral acids. Of course all these species have their own redox behaviour; values of the reduction potentials which have been determined or estimated for these moieties are shown in equations (14)–(19).

$$
\begin{array}{lcr}
 & E^0/V & \\
A + 2H^+ + 2e \rightleftharpoons H_2A & 0.40 & (14) \\
A + H^+ + 2e \rightleftharpoons HA^- & 0.28 & (15) \\
A + 2e \rightleftharpoons A^{2-} & -0.05 & (16) \\
A + e \rightleftharpoons A^- & 0.16 & (17) \\
A^- + e \rightleftharpoons A^{2-} & +0.05 & (18) \\
HA^\cdot + e \rightleftharpoons HA^- & +0.70 & (19) \\
\end{array}
$$

Any study of the mechanism of a reaction in aqueous solution will usually involve a stage where there is an effect of pH on the nature of the reactants and therefore on the rate of the reaction. This usually provides a valuable insight into the nature of the reaction and the way it proceeds.

We will see later that many studies have been carried out on reactions between L-ascorbic acid and transition metal complexes and in virtually all cases a complex rate law has had to be devised in order to account for the effect of hydrogen ion concentration on the rate of the reaction.

REDOX REACTIONS OF VITAMIN C

The complexity of the redox processes involving vitamin C have not prevented numerous studies of the oxidation of L-ascorbic acid and some on the oxidation and reduction of dehydroascorbic acid and other oxidation products. One of the reasons for the great interest in this subject is that the role of vitamin C in living systems is certainly connected to its oxidation–reduction behaviour and this may ultimately be the key to understanding the biological mechanisms of the actions of vitamin C. In this section we will examine the oxidation and reduction of vitamin C by a wide variety of reagents, with emphasis on the behaviour of L-ascorbic acid since this is the most studied compound in this respect.

Oxidation of L-Ascorbic Acid by Dioxygen

Even before the nature of the 'antiscorbutic factor' was known it was realised that the compound behind it was susceptible to atmospheric oxidation. L-Ascorbic acid is oxidised by dioxygen to dehydroascorbic acid, but in one sense this does not really matter, since both have the antiscorbutic and other physiological effects of vitamin C. *In vivo* the reaction of L-ascorbic acid to dehydroascorbic acid may to all intents and purposes be regarded as reversible. However, for the study of the properties of L-ascorbic acid *in vitro*, it is clear that it is always important to use freshly prepared solutions which have been made up in oxygen-free water. It is apparent that dioxygen produces further damage to vitamin C by oxidising dehydroascorbic acid to products which are no longer physiologically active. This process occurs in aqueous solution exposed to air and it is enhanced by the intervention of catalysts when the vitamin C is present in fruit and vegetables. It is believed that enzymic action is largely responsible for this destruction of vitamin C in air. Similarly many transition metal ions catalyse both the oxidation of L-ascorbic acid to dehydroascorbic acid and also the destruction of the dehydroascorbic acid in air to 2,3-diketogulonic acid and other oxidation products. Copper(II) ions are particularly active in this respect and catalysis by these ions is thought to be responsible for the loss of vitamin C from orange, lemon, and lime juice concentrated in copper vessels in the latter half of the nineteenth century. This caused a loss of faith in the efficacy of fresh fruit in the prevention and cure of scurvy and probably set back the discovery of the active agent in these fruits by a number of years.

The oxidation of L-ascorbic acid by dioxygen dissolved in water has been the subject of a large number of studies. Some have concentrated on the mechanism of the reaction which occurs in solutions from which transition metal ions have been rigorously excluded, while others have sought to find out the role of transition metals in catalysing the reaction. The effects of copper(II) ions in particular have been prominent in these studies.

Oxidation in the Absence of Catalysts

The term 'auto-oxidation' is often used to refer to the oxidation of L-ascorbic acid by molecular oxygen in aqueous solution in the absence of any catalysts. The many studies on this topic still have not produced an entirely satisfactory mechanism for the electron transfer from ascorbate to the dioxygen molecule. E.s.r. studies have shown that when ascorbate reacts with dioxygen in aqueous solution in the pH range 6.6–9.6, a steady-state concentration of ascorbate radicals is produced. There is, however, little direct evidence for the formation of H_2O_2 in this reaction. Data from the inhibition of the reaction by certain enzymes, notably superoxide dismutase, suggest that H_2O_2 is produced in the process.

In the absence of catalysts the reaction of L-ascorbic acid with dioxygen is slow, though the rate is very pH dependent. It is slowest in acid solution, and at pH 4 a typical rate constant is 10^{-5} l mol^{-1} s^{-1}. Even at pH 10, although the rate is much higher, it is still only about 5 l mol^{-1} s^{-1}. This has been suggested as a possible explanation for the absence of any direct evidence for the presence of O_2, since it would only be present in the smallest concentrations at any time during such slow reactions.

Oxidation in the Presence of Metal Ions

Oxidation of L-ascorbic acid itself, dehydroascorbic acid, and the further oxidation products appears to be subject to metal ion catalysis of which copper(II) is the most potent. There have been many investigations of the reactions of L-ascorbic acid under these conditions, but very few of the catalytic oxidation of the other compounds. Many of the studies have sought to establish the role of the metal ion, usually copper(II) in the mechanism of the reaction. Although the picture is somewhat clearer today, there remains disagreement about the role of the metal ions in the process.

The important work of A. E. Martell in this field showed that, under the conditions he used, the reaction was first-order in metal ion, in ascorbic acid, and in dioxygen concentration. The mechanism proposed for the reaction involves the formation of a ternary complex between L-ascorbic acid, copper(II), and dioxygen, which is shown in Figure 7.12. This is not the end of the story, however. It is known that the order with respect to dioxygen decreases as the concentration of dioxygen decreases. It is against this background that more recently it has been shown that the order with respect to dioxygen at low concentrations is one half, so that the rate law for the disappearance of dioxygen is then:

$$\frac{-d[O_2]}{dt} = k[Cu^{2+}][HA^-][O_2]^{\frac{1}{2}} \qquad (20)$$

In order to explain the half-order dependence on dioxygen concentration it has been found necessary to propose rather exotic complexes between copper(II) and L-ascorbic acid:

$$2[CuHA]^+ \rightleftharpoons [CuHA]_2^{2+} \qquad (21)$$

and a suggested structure for this complex is given in Figure 7.13. This complex interacts with dioxygen to produce a further intermediate:

$$[Cu(HA)]_2^{2+} + O_2 \rightleftharpoons [CuHAO_2CuHA]^{2+} \qquad (22)$$

The suggested structure for this dioxygen complex is that shown in Figure 7.14. This latter may then decompose in a rate determining step:

$$[CuHAO_2CuHA]^{2+} \rightarrow ACuO_2H\cdot + Cu^{2+} + A\cdot \rightarrow product \qquad (23)$$

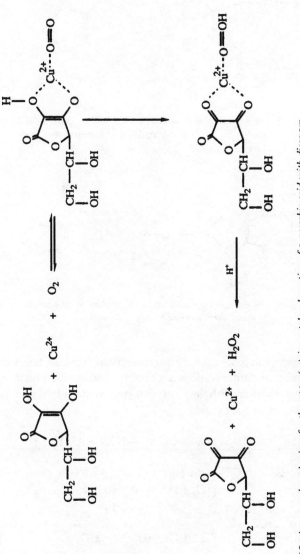

Figure 7.12 *A proposed mechanism for the copper(II) ion catalysed reaction of L-ascorbic acid with dioxygen*

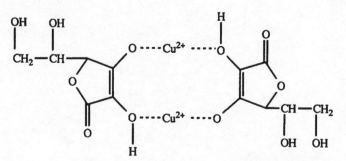

Figure 7.13 *Intermediate proposed to explain a half-order dependence on dioxygen concentration in the rate of oxidation of L-ascorbic acid in the presence of copper(II) ions*

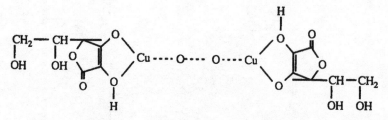

Figure 7.14 *The dioxygen molecule complex intermediate in the oxidation of L-ascorbic acid catalysed by copper(II) ions*

However, this is not the only interpretation of this rate law which has been proposed. Other workers have interpreted similar data in terms of a chain mechanism with the formation and reaction of copper(I) ions as a key feature in the process:

Initiation	$2Cu^{2+} + HA^- \rightarrow 2Cu^+ + A + H^+$	(24)
Propagation	$H^+ + Cu^+ + O_2 \rightarrow Cu^{2+} + HO_2$	(25)
	$HO_2 + HA^- \rightarrow A^- + H_2O_2$	(26)
	$A^- + Cu^{2+} \rightarrow Cu^+ + A$	(27)
Termination	$Cu^+ + HO_2 \rightarrow Cu^{2+} + HO_2$	(16)

No doubt the truth lies somewhere between these two rather different mechanisms, but it does show how the same or similar data can give rise to quite different interpretations. There is no doubt that,

Table 7.1 *Some equilibrium constants (K) for ascorbic acid/metal ion complexes*

Metal ion	Equilibrium	25 °C, 1.0 M log K (ionic strength)
Mn^{2+}	MHL/M.HL	1.1 (0)
Fe^{2+}	MHL/M.HL	0.21 (1.0)
	ML/M.L	1.99 (1.0)
Co^{2+}	MHL/M.HL	1.4 (0)
Ni^{2+}	MHL/M.HL	1.1 (0)
Cu^{2+}	MHL/M.HL	1.57 (0.1)
VO_2^{+}	$MH_2L/M.H_2L$	2.69
Ag^{+}	ML/M.L	3.66 (0.1)
Zn^{2+}	MHL/M.HL	1.0 (0)
Pb^{2+}	MHL/M.HL	1.77 (0.1)
Au^{3+}	MHL/M.HL	1.89 (0.1)
	$M(HL)_2/M.(HL)_2$	3.55 (0.1)

in an aqueous solution of copper(II) ions in L-ascorbic acid, there is a rapid equilibrium which results in the formation of the complex $Cu[HA]^{+}$ and that substantial amounts are present in solutions containing a large excess of L-ascorbic acid over copper(II). The system becomes even more complicated when chelating agents such as EDTA or other complexing agents such as chloride or bromide ions are present. In the latter case a significant acceleration of the reaction is observed at low concentration (<0.05M). This has been variously interpreted, but is also a feature of other copper(II)-catalysed reactions of L-ascorbic acid.

Reactions with Metal Ions

Vitamin C forms complexes with metals, even those which it is capable of reducing, such as iron(III) and copper(II). It is a potentially bidentate ligand. The X-ray structure of only one complex has been determined and this will be discussed later. In general, it is assumed that most complexes will be formed by co-ordination to the oxygens of the 2- and 3-hydroxides. If this is the case then, as has been pointed out by Martell, the complexes formed are weaker than they should be in comparison to the complexes of similar chelating ligands. Table 7.1 shows a selection of formation constants for complexes with a number of different metal ions. It can be seen that the highest value is 3.66, which is well below the value of 5 expected by comparison with similar ligands. The complexes shown in Table 7.1 are all with HA^{-}. There are very few, if any, complexes recorded with the fully deprotonated ligand. There may be many reasons for this, but it is likely that because such complexes are formed in alkaline solution, many of the metal ions which would be expected to form complexes produce hydroxo species in aqueous solution.

When a solution of iron(III) chloride is reacted in slightly alkaline solution with L-ascorbic acid, there is an immediate deep-blue colour produced and from this solution a dark blue, almost black powder may readily be isolated. In acid solution this color appears as a transient 'flash' of blue when the two solutions are mixed and before the iron(III) is reduced to iron(II) by the L-ascorbic acid. The structure of the dark-blue complex has not been determined conclusively, but it is clear that it is a complex of ascorbic acid with iron. Figure 7.15 shows a structure which has been suggested for this complex.

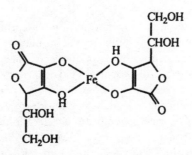

Figure 7.15 *Suggested structure of the L-ascorbate iron(III) complex*

The structures of such complexes in aqueous solution are largely unknown. There have been n.m.r. studies of complexes such as the nickel(II) complex and the structure shown in Figure 7.16 has been deduced from these data.

Recently crystals of a number of complexes of L-ascorbic acid were obtained by Jabs and Gaube using reactions of the type:

$$NiSO_4.7H_2O + Ba(OH)_2 + 2HA^- \rightarrow Ni(HA)_2.4H_2O + BaSO_4 + 3H_2O$$
$$(29)$$

The key to the success of this process is that it is carried out under strictly anaerobic conditions. Complexes which have been prepared by this technique are: $TiO(HA)_2.2H_2O$, $TiO(OH)(HA)$, $Cr(HA)_3.6H_2O$, $Mn(HA)_2.4H_2O$, $Co(HA)_2.4H_2O$, $Ni(HA)_2.4H_2O$, $Zn(HA)_2.4H_2O$.

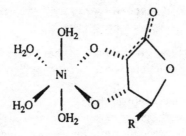

Figure 7.16 *Structure of the nickel(II) complex with L-ascorbic acid proposed on the basis of n.m.r. studies*

At the time of writing the X-ray crystal structures of these complexes have not been determined.

The only X-ray crystal structure of a transition metal complex of L-ascorbic acid which has been determined to date is that of *cis*-1,2-diaminocyclohexane(ascorbato)platinum(II). Unexpectedly, this compound was not found to involve complexation through the 2-OH and 3-OH of the ascorbic acid, but via the C-2 and the deprotonated hydroxyl in the 5-position! The structure is shown in Figure 7.17.

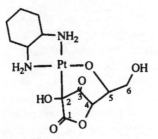

Figure 7.17 *The* cis *dach platinum(II) complex of L-ascorbic acid*

Although at first sight the structure may appear to throw into question the structures of all other complexes between L-ascorbic acid and transition metal ions, it should be noted that platinum(II) sometimes shows a preference for the formation of Pt–C bonds where Pt–O bonds might have been expected [*cf.* the complex between platinum(II) and pentane-2,4-dione, acetylacetone]. A similar structure was proposed for the bis(ascorbate) complex on the basis of the n.m.r. spectrum. The *cis* dach complex and similar complexes have shown some promise as anti-tumour agents.

Oxidation of L-Ascorbic Acid by Transition Metal Complexes

Redox reactions of transition metal ions have been studied for many years. Broadly the reactions which occur may be classified into two types:

(1) Inner-sphere electron transfer:

The transfer of the electron is believed to occur by the formation of a 'bridge' between the reductant and oxidant. This provides a route for the electron to be transferred from one to the other. The bridged species is the transition state of the reaction.

(2) Outer-sphere electron transfer:

In these the transfer of the electron from the reducing agent to the metal occurs without any bond being formed between the reductant and oxidant.

It is found that the reduction of transition metal ions and complexes by L-ascorbic acid occurs by both the outer- and inner-sphere mechanisms depending on the nature of the complex or the metal. Many examples of both routes have been studied.

The role of transition metals in the catalysis of redox reactions of L-ascorbic acid frequently involves a step in which the transition metal ion is itself reduced.

Because we live and carry out our chemistry in a world in which water is ubiquitous, we tend to forget that transition metal ions in aqueous solutions are present as aqua complexes and should not be regarded as 'free ions'. In aqueous solution many transition metal ions are reduced by L-ascorbic acid. A typical example in which the mechanism of the reaction has been studied is that with cobalt(III) ions in water. Cobalt(III) ions have only a comparatively brief existence in water, but exist long enough to allow kinetic experiments to be performed on their reaction with L-ascorbic acid. The high positive charge on the cobalt causes the water molecules in the aqua complex to be acidic, so that hydroxo species are formed at higher pH values. The effect of this on the reduction reaction is reflected in the mechanism shown in equations (30)–(34).

$$Co^{3+}(aq) \rightleftharpoons [CoOH]^{2+}(aq) + H^+ \tag{30}$$

$$Co^{3+} + H_2A \rightarrow Co^{2+} + H_2A^+ \tag{31}$$

$$[Co(OH)]^{2+}(aq) + H_2A \rightarrow Co^{2+}(aq) + HA\cdot + H_2O \tag{32}$$

$$Co^{3+}(aq) + HA\cdot \rightarrow Co^{2+}(aq) + A + H^+ \tag{33}$$

$$[CoOH]^{2+}(aq) + HA\cdot \rightarrow Co^{2+}(aq) + A \tag{34}$$

A similar mechanism has been demonstrated for other aqua ions such as aquamanganese(III) ions, but here it was possible to extend the pH range to include HA^- as a reactant.

These reactions illustrate a common thread which runs through the mechanisms of most oxidations of L-ascorbic acid by transition metal ions and complexes, *i.e.* the first oxidation step invariably results in the formation of an ascorbate radical. Reaction of that radical with another metal ion is commonly proposed and this gives the required stoichiometry.

Many complexes other than the aqua species have been studied. A typical example of these complexes is the reduction over a range of pH of the anion tris(1,2-ethanedioato)cobaltate(III), $[Co(C_2O_4)_3]^{3-}$. A proposed mechanism is shown in equations (35)–(39).

$$H_2A \underset{}{\overset{K_a}{\rightleftharpoons}} HA^- + H^+ \tag{35}$$

$$[Co(C_2O_4)_3]^{3-} + H_2A \overset{k_1}{\rightarrow} H_2A^+ + Co^{2+} + 3[C_2O_4]^{2-} \tag{36}$$

$$[Co(C_2O_4)_3]^{3-} + HA^- \overset{k_2}{\rightarrow} HA\cdot + Co^{2+} + 3[C_2O_4]^{2-} \tag{37}$$

$$[Co(C_2O_4)_3]^{3-} + H_2A^+ \rightarrow DHA + Co^{2+} + 3[C_2O_4]^{2-} + 2H^+ \tag{38}$$

$$[Co(C_2O_4)_3]^{3-} + HA\cdot \rightarrow DHA + Co^{2+} + 3[C_2O_4]^{2-} + H^+ \tag{39}$$

This gives the rate law:

$$k_{obs} = k_1 + (k_2 - k_1) \frac{K_a}{([H^+] + K_a)} \tag{40}$$

Similar rate laws are obtained for many other reactions and the oxidation mechanism has been studied for complexes of virtually all the metal ions which are capable of oxidising L-ascorbic acid. Sometimes fairly exotic species are used, such as the complex of the nickel(IV) ion with diacetyloxime triethylenetetramine Schiff's base (Figure 7.18), which gives the complex shown in Figure 7.19.

In this case the reaction is a two-electron redox process, since the product is nickel(II). E.s.r. studies of the reaction suggest that it occurs

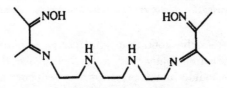

Figure 7.18 *Diacetyloxime triethylenetetramine Schiff's base*

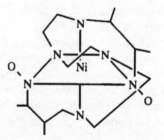

Figure 7.19 *Skeleton structure of the nickel(IV) complex with 2,4-dimethyl-4,7,10,13-tetra-azahexadecane-3,13-dione dioxime*

in two discrete one-electron steps. The pH behaviour this time is complicated by the protonation of the complex, which affects its redox properties.

Like the oxidation by dioxygen, these reactions are found to be catalysed by copper(II) ions, although such reactions have not been widely studied. The reduction of tris(1,2-ethanedioato)cobaltate(III) is strongly catalysed by copper(II) and the kinetics of this reaction have been interpreted as involving a pathway in which the copper(II) ions act as an intermediary for the passage of the electron from the L-ascorbic acid to the cobalt(III), via an intermediate postulated to be that in Figure 7.20. Such reactions have been shown to be subject to the same accelerating effect of small concentrations of halide ions as the dioxygen oxidation.

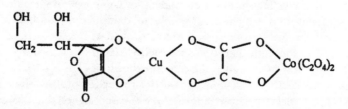

Figure 7.20 *Possible structure of the intermediate in the copper(II) catalysed oxidation of L-ascorbic acid by the tris(oxalato)cobaltate(III) ion*

Clock reactions give good results in lecture-demonstrations. They have been known for many years and often involve the reaction of bromate with iodide in the presence of a reducing agent, and L-ascorbic acid has been used as the reducing agent in such reactions. Metal ions have been found to catalyse this reaction as well.

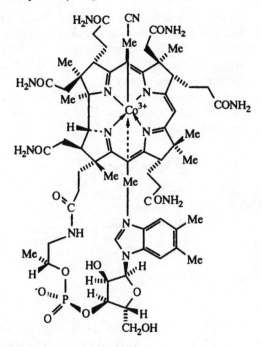

Figure 7.21 *Vitamin B₁₂, cyanocobalamin*

Molybdenum(VI) increases the rate of the reaction between bromate and iodide, while vanadium(V) catalyses the direct reaction between bromate and L-ascorbic acid. The effect of this is to reduce the induction period of the reaction.

Reaction of Vitamin C with Vitamin B₁₂

Vitamin B_{12} is a cobalt complex. It is essential for good health in man and without it pernicious anaemia will eventually result in death. The vitamin is not synthesised in the human body and has to be obtained from food, mostly meat and dairy products, or micro-organisms. It is the only vitamin which contains a metal ion and furthermore it is often found *in vivo* with a cobalt–carbon bond – one of Nature's rare organometallic compounds. It is usually used in the laboratory as the beautiful red crystalline solid, which is the cyano-complex. The structure of vitamin B_{12} was originally determined using this compound, cyanocobalamin, and this is shown in Figure 7.21. The cyano-species is only found in the bodies of people who smoke! The cyanide is replaced by a variety of ligands in the different forms of the vitamin and it is this axial bond, above the plane of the corrin ring, which is frequently involved in the biochemical role of the vitamin. The structure given in Figure 7.21 has been determined by *X*-ray diffraction and has also been confirmed by chemical means. Perhaps the most significant feature of the structure is the planar corrin ring, at the centre of which is the cobalt ion. This ring (Figure 7.22) is essentially a porphyrin ring with C-20 missing.

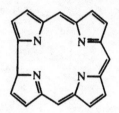

Figure 7.22 *The corrin ring structure*

The sixth position is occupied by a 5,6-dimethylbenzimidazole grouping (Figure 7.23) below the plane of the ring. The absorption of this vitamin by the body depends upon the presence of a mucoprotein called the 'intrinsic factor'. This binds to the vitamin B_{12} in a stoichiometric ratio. The total body pool of the vitamin is quite small, in the region of 3–5 mg, and most of this is stored in the liver in the deoxyadenosyl form. The daily requirements in man are only about 3–7 μg. This amount is present in the normal diet, but may not be in some strict vegetarian diets, since vitamin B_{12} is only found in some plant unicells, *e.g. Euglena*.

The cyanide ligand may be replaced by water to give the so-called aquacobalamin, or vitamin B_{12a}. In these compounds the cobalt is

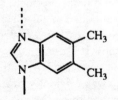

Figure 7.23 *The 5,6-dimethylbenzimidazole part of vitamin B_{12}*

present in the +3 oxidation state. In the two biologically active coenzyme forms, the axial position is occupied by methyl (methylcobalamin) or, as coenzyme B_{12}, by 5'-deoxyadenosyl (Figure 7.24). Both of these forms of the vitamin are extremely light-sensitive.

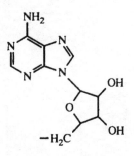

Figure 7.24 *The 5'-deoxyadenosyl group*

Fairly mild reduction of cyano- or aqua-cobalamin reduces the cobalt(III) to cobalt(II) (usually yellow/orange). Further reduction with more powerful reducing agents results in cobalt(I) species (usually green or grey). This may also be accompanied by various changes in the nature of the co-ordination around the cobalt ion, though the corrin ring structure remains undisturbed. This is shown schematically in Figure 7.25.

Fears have been voiced that vitamin C will destroy vitamin B_{12} and that a person taking 'megadoses' of vitamin C (about 5–10 g per day) could possibly be in danger of destruction of the body stores of vitamin B_{12} and hence be liable to pernicious anaemia. In fact it has been found that addition of vitamin C in large quantities to food and incubation at 37 °C had no effect upon the vitamin B_{12} content. Similarly, experiments with rats have shown that feeding with vitamin C did not deplete the vitamin B_{12} contents of the plasma or liver. No vitamin B_{12} depletion was observed in male patients suffering from spinal cord injury, when 4 g of vitamin C per day was given over eleven months.

However, there is no doubt that Vitamin C and vitamin B_{12} react with each other *in vitro*. Furthermore, it is known that where reaction does not occur in the absence of copper(II) (as in the case of cyanocobalamin) it does happen when copper(II) ions are present. Also copper(II) ions have a great accelerating effect on reaction between the two vitamins.

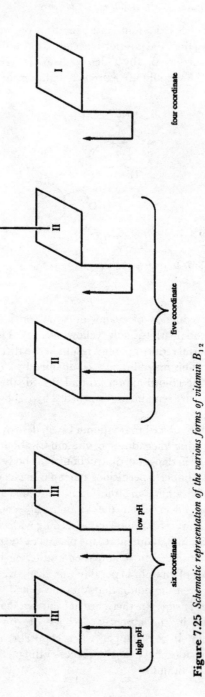

Figure 7.25 *Schematic representation of the various forms of vitamin B_{12}*

Experiments have shown that when L-ascorbic acid is reacted with cyanocobalamin in the presence of copper(II) ions in aqueous solution, the red colour of the cyanocobalt(III) species is changed into the orange colour of the aquacobalt(II) compound. Bubbling air through this solution or the addition of aerated water quickly restores the red colour. On leaving this solution, the colour changes back to orange. This may be repeated as long as there is any L-ascorbic acid remaining. Aquacobalamin (vitamin B_{12a}) reacts readily with L-ascorbic acid in the absence of copper(II). This reaction is very sensitive to the presence of oxygen, which has to be rigorously excluded. Once more the product is the cobalt(II) compound. The rate of the reaction is pH-dependent. As the pH is raised from 2.0, the rate increases until it reaches a maximum at about pH 5.6. Beyond this the rate decreases, reaching a constant value at about pH 8.5. Above pH 2.0 the HA^- ion is produced with the usual increase in rate. The rate levels off when virtually all the ascorbic acid is present as the monoanion and then begins to decrease again at higher pH as the water on the aquacobalamin loses a proton. The resulting hydroxocobalamin is reduced more slowly than the aquacobalamin.

Reduction by the Inner-sphere Mechanism

This mechanism involves the formation of a chemical bond between the two reacting species, which may act as a 'bridge' for the transfer of the electron. This is the less common method of electron transfer in redox reactions involving L-ascorbic acid. We have already seen that reaction of iron(III) results in the formation of an intermediate blue compound and it is likely that this is an inner-sphere electron transfer.

The kinetics of the reaction of vanadium(V) with L-ascorbic acid have been studied and the rapid formation of a brown intermediate with a visible absorption maximum at 425 nm is seen. The interpretation of the data suggests the mechanism shown in equations (41)–(43).

$$V^V + H_2A \rightleftharpoons [V^V H_2A] \qquad (41)$$
$$[V^V H_2A] \rightarrow V^V + \text{radical} \qquad (42)$$
$$\text{radical} + V^V \rightarrow V^{IV} + A \qquad (43)$$

This reaction is unusual, in that complex formation occurs even at comparatively high acid concentrations, when H_2A is the predominant species present.

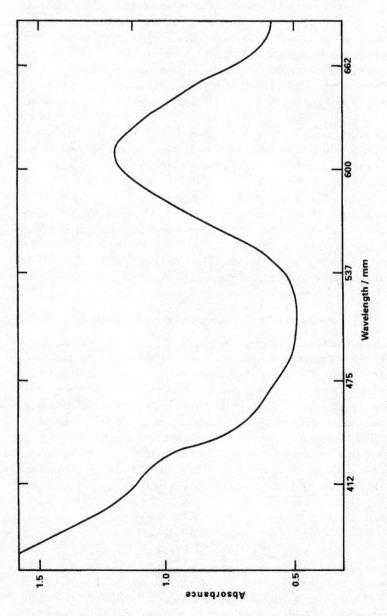

Figure 7.26 *Absorption spectrum of ascorbate oxidase*

ASCORBATE OXIDASE

No discussion of the inorganic chemistry of vitamin C would be complete without some mention of the enzyme ascorbate oxidase. This enzyme catalyses the oxidation of L-ascorbic acid to dehydroascorbic acid. It was originally discovered in cabbage leaves by Szent-Györgyi, the first person to isolate vitamin C. When the enzyme is obtained from summer crookneck squash, it is blue–green and has a relative molar mass of about 150 000. It is a copper enzyme and contains about 0.26% copper. The absorption spectrum is shown in Figure 7.26. The most prominent feature is the peak at 610 nm, which is attributed to copper(II) in the enzyme. There is also a shoulder at about 412 nm. The enzyme is readily bleached to light yellow by the addition of L-ascorbic acid at about pH 6. This process is due to the conversion of the copper(II) into copper(I) and the blue colour is restored by exposing the mixture to oxygen.

Much has been written about the mode of action of ascorbate oxidase and other oxidases, but until 1989 the structure of the former was unknown. In that year Albrecht Messerschmidt and co-workers crystallised two different modifications of the oxidase, both having blue crystals and determined the X-ray crystal structure of ascorbate oxidase from zucchini. Two crystal forms were studied and the crystal structure of each was obtained. One crystal form was a dimer with a relative molar mass of 140 000 and the other a tetramer with a relative molar mass of 280 000. The nature of the copper ions in enzymes of this sort has often been a matter of contention before the crystal structure has been determined. This has also been the case for ascorbate oxidase. The X-ray analysis showed that each subunit had four copper atoms which are arranged so that there is one 'cluster' of three atoms, associated with which are eight histidine ligands, and a mononuclear copper atom which has a cysteine, a methionine, and two histidine ligands associated with it. There is also evidence that the trinuclear cluster consists of two copper atoms of a similar type, whereas the third has a different function. A drawing of the dimer of ascorbate oxidase is shown in Figure 7.27. There is no doubt that the determination of the crystal structure of this enzyme will be an important landmark in the understanding the mechanisms of the actions of these oxidases.

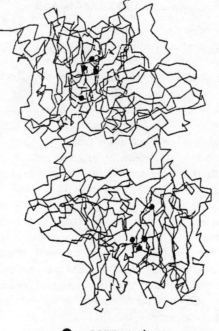

● = copper atom

Figure 7.27 *A drawing of the dimer of ascorbate oxidase*

Bibliography

The following are texts which could be used for further reading on vitamin C and related subjects. This is not intended (nor could it be) as an exhaustive list of publications on vitamin C. There is a vast primary literature on all aspects of vitamin C chemistry and biochemistry and also many reviews on the topic. Some of the books listed here will provide detailed references which will allow the reader to get into that literature if required.

T. K. Basu and C. J. Schorah, 'Vitamin C in Health and Disease', Croom Helm, London and Canberra, 1982.

J. J. Burns, J. M. Rivers, and L. J. Machlin, 'Third World Conference on Vitamin C', New York Academy of Sciences, 1987.

K. J. Carpenter, 'The History of Vitamin C and Scurvy', Cambridge University Press, Cambridge, 1986.

E. Cheraskin, N. M. Ringsdorf Jr, and E. L. Sisley, 'The Vitamin C Connection' Thorsons Publishers, Wellingborough, 1983.

J. J. Chinoy, 'The Role of Ascorbic Acid in Growth, Differentiation and Metabolism of Plants', Kluer Boston, The Hague, 1987.

T. P. Coultate, 'Food: The Chemistry of its Components', 2nd Edn., Royal Society of Chemistry, London, 1989.

J. N. Counsell, and D. H. Hornig, 'Vitamin C, Ascorbic Acid', Applied Science, London, 1981.

R. W. Hay, 'Bio-Inorganic Chemistry', Ellis Horwood, Chichester, 1984.

S. M. Kybett, 'Henry VIII – A Malnourished King', *History Today*, 1989, **39**, 19–25.

M. Levine, 'New Concepts in the Biology and Biochemistry of Vitamin C', *New Engl. J. Med.*, 1986, **314**, 892–902.

S. Lewin, 'Vitamin C, its Molecular Biology and Medical Potential', Academic Press, 1976.

R. W. Moss, 'Free Radical: Albert Szent-Györgyi and the Battle over Vitamin C', Paragon House, New York, 1988.

E. A. Newsholme and A. R. Leech, 'Biochemistry for the Medical Sciences', Wiley, Chichester, 1983.

S. Nobile and J. M. Woodhill, 'Vitamin C. The Mysterious Redox System, A Trigger of Life', MTP Press, Lancaster, 1981.

L. Pauling, 'Vitamin C and the Common Cold', Ballantine Books, London, 1970.

W. H. Sebrell and R. S. Harris, 'The Vitamins', Academic Press, New York and London, 1967.

P. A. Seib and B. M. Tolbert, 'Ascorbic Acid: Chemistry, Metabolism, and Uses', Advances in Chemistry Series, American Chemical Society, Washington DC, 1982.

M. R. Werbach, 'Nutritional Influences on Illness', Thorsen Publishing Co. Wellingborough, 1989.

J. B. Wyngaarden and L. H. Smith, 'Cecil Textbook of Medicine', 18th Edn., Saunders Co., 1988.

The key references on the first isolation, structure determination and synthesis of vitamin C are:

Isolation: A. Szent-Györgyi, *Biochem. J.*, 1928, **22**, 1387.

Structure determination: R. W. Herbert, E. L. Hirst, E. G. V. Percival, R. J. W. Reynolds, and F. Smith, *J. Chem. Soc.*, 1933, 1270.

Synthesis (from L-xylosome): W. N. Haworth and E. L. Hirst, *Chem. Ind. (London)*, 1933, 645.

R. G. Ault, D. K. Baird, H. C. Carrington, W. N. Haworth, R. W. Herbert, E. L. Hirst, E. G. V. Percival, F. Smith, and M. Stacey, *J. Chem. Soc.*, 1933, 1419.

Synthesis (from D-glucose): T. Reichstein and A. Grüssner, *Helv. Chim. Acta*, 1934, **17**, 311.

Subject Index